# BIOFABRICATION

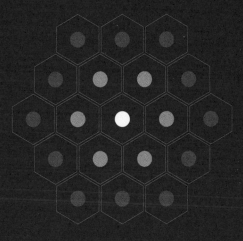

**The MIT Press Essential Knowledge series**

A complete list of the titles in this series appears at the back of this book.

# BIOFABRICATION

RITU RAMAN

The MIT Press | Cambridge, Massachusetts | London, England

The MIT Press would like to thank the anonymous peer reviewers who provided comments on drafts of this book. The generous work of academic experts is essential for establishing the authority and quality of our publications. We acknowledge with gratitude the contributions of these otherwise uncredited readers.

This book was set in Chaparral Pro by New Best-set Typesetters Ltd. Printed and bound in the United States of America.

Library of Congress Cataloging-in-Publication Data

Names: Raman, Ritu, author.
Title: Biofabrication / Ritu Raman.
Description: Cambridge, Massachusetts : The MIT Press, [2021] | Series: The MIT Press essential knowledge series | Includes bibliographical references and index.
Identifiers: LCCN 2020047113 | ISBN 9780262542968 (paperback)
Subjects: LCSH: Tissues—Models. | Biomedical engineering—Moral and ethical aspects. | Microfabrication.
Classification: LCC R857.T55 R36 2021 | DDC 610.28—dc23
LC record available at https://lccn.loc.gov/2020047113

10   9   8   7   6   5   4   3   2   1

In gratitude for the three engineers who have inspired me throughout my life:

Radha Raman, Raman Radhakrishnan, and K. R. Dorairaj.

# CONTENTS

# SERIES FOREWORD

The MIT Press Essential Knowledge series offers accessible, concise, beautifully produced pocket-size books on topics of current interest. Written by leading thinkers, the books in this series deliver expert overviews of subjects that range from the cultural and the historical to the scientific and the technical.

In today's era of instant information gratification, we have ready access to opinions, rationalizations, and superficial descriptions. Much harder to come by is the foundational knowledge that informs a principled understanding of the world. Essential Knowledge books fill that need. Synthesizing specialized subject matter for nonspecialists and engaging critical topics through fundamentals, each of these compact volumes offers readers a point of access to complex ideas.

Biofabrication, the act of building with biological materials, may sound too technologically complex to be easily accessible. Our lived experiences in the natural world, however, give us tremendous insight into the behavior of biological materials. Each of us already has the baseline intuition we need to understand the scope and impact of this emerging scientific discipline.

Imagine yourself starting a vegetable patch in a community garden. You carry a large bag of tools, soil, seeds, and a watering can to the garden a mile away from your home and establish a small shaded plot for growing a crop of carrots. Every day, you jog to the garden to check on your carrots, water them, and run home. As the first leaves on your plants start to sprout, you notice a family of rabbits poking around your vegetable patch and sprinkle the leaves with red pepper. As time passes, your legs get stronger running to the garden every day, the carrot leaves grow larger and turn to face the sun, and the rabbits have learned not to eat your plants. This is the power of biological materials.

The living cells that make up your body, and the plants and animals that surround us, are immensely powerful. They can sense a range of external signals and adapt their form and function to best suit their environment. These behaviors manifest in many different ways, from changes in the shape of your body to changes in the learned behavior of animals, as illustrated in the example above. The synthetic world that we have built around us, while vastly complex and powerful in its own right, is still far from being able to match the responsive nature of our natural world. Our clothes do not adapt to changing weather nor do our cars become larger to accommodate a growing family. Why not?

As a mechanical engineer, I was trained to build with traditional synthetic materials, as engineers have for centuries. For most engineers, building things from the ground up using metals and plastics seems more natural, oddly enough, than building with naturally existing biological materials. Luckily, I was raised by a family of engineers who taught me to be observant of my surroundings, and I have had the privilege to pursue science education in an era when biology is converging with engineering. This intellectual environment, combined with my first job as an undergraduate at Cornell University, shifted my perspective on the materials with which engineers can (and should) build.

My first job as an undergraduate researcher was to contribute to a Cornell lab's research on the effect of diet and exercise on the composition and organization of skeletal muscle. This is an elaborate way of saying that my job was to feed rats different diets, observe them running on tiny treadmills, and record the changes in their muscles. This was perhaps the first time during my academic training that I felt my experiences at school correlating to my experiences in life. I had just started running regularly and had never stopped to think about the fact that my body was changing at a cellular level to make me stronger and faster. Researching the complex mechanisms by which the rats' muscles bulked up in times of need made me all the more appreciative of how regular exercise was shaping my own body. This experience made me realize, for the first time, that I was a machine. I was a biological machine whose movement was powered by skeletal muscle, just as a car is a machine whose movement is powered by an engine. But, unlike a car, I could make myself stronger when needed. I could do this because I was composed of biological materials, and this made me think that perhaps other machines should be built with biology as well.

This idea is the core of biofabrication. Building with biology, in the same way we build with synthetic materials, is a novel discipline that is rapidly changing. It has potential applications in medicine, agriculture, robotics, and beyond, and cultivating literacy around the fundamentals of this field is critical to understanding and shaping the future technological landscape of our world. While biofabrication is broadly defined as building with biological materials, which can include nucleic acids and proteins and cells of all kinds, this book focuses on biofabrication in the context of building with living cells from mammals. This focus allows us to take a closer look at some of the most impactful applications of biofabrication while still giving you the vocabulary you need to understand other forms of biofabrication you may choose to explore in the future.

Chapters 1 and 2 cover the essential knowledge required to understand the emergence and evolution of biofabrication, as well as the fundamental technology that enables building with living cells. Chapters 3–6 take deep dives into four applications of biofabrication that are most likely to impact our daily lives in the coming years—namely, tissue engineering, organs-on-a-chip, lab-grown meat and leather, and biohybrid machines. Chapters 7–9 present a discussion of the environmental, economic, and ethical implications of biofabrication and propose a future scientific and regulatory vision for this field and our global community.

This book will train you on the fundamentals of biofabrication and serve as a guide for conducting further independent exploration of the topics presented, as well as of the field more broadly. Combining this new knowledge with your own lived experience in the natural world, which has given you an intuitive understanding of how biological systems sense and adapt to their environment, will open up new possibilities for the world of tomorrow. If nothing else, I hope this short introduction to biofabrication makes you stop and appreciate the beauty, adaptiveness, and persistence of the biological machinery that drives your body and our world.

# INTRODUCTION

Biological systems sense and respond to their surroundings. When you exercise, you get stronger. When you cut your skin, you heal. The built environment and the machines that surround us, however, do not have such abilities. Why not? The answer is quite simple: they are not built out of the living biological materials that form our bodies and our natural organic environment.

Throughout human history, we have built devices with materials such as stone, metal, or plastics that perform the same function over and over again until they are damaged or worn away. We have optimized these materials for strength, durability, and reduced cost and built a fantastic array of machines that help us live safer, more comfortable, and more productive lives. Recent progress in developing new smart materials out of composite polymers and complex electronic materials out of semiconductors has even helped make devices that can sense and respond to certain types of external stimuli. A smartphone, for example, can sense the ambient environmental light and dim its screen accordingly to maximize user comfort. However, we are still very far from being able to fully replicate the abilities of biological materials and the way they continually adjust their form and function to match changing needs. A scratched phone screen remains scratched, often deteriorating further with time. Scraping the skin on our bodies, however, triggers a complex cascade of microscopic and macroscopic events that alert us to damage and heal the cut, autonomously returning the skin to its native unharmed state. No synthetic material or machine we have built to date can match the level of immense complexity and innate responsiveness observed in biological systems.

Why does this matter? Some of the biggest scientific challenges facing our world, such as fighting deadly diseases and generating sustainable sources of food and energy, are incredibly dynamic. Moreover, the global population and corresponding need for critical medical and agricultural resources are rapidly growing. The static materials with which we have been building for millennia cannot be the sole solution to these dynamic problems. This motivates adding a new set of materials into the toolbox of every inventor by teaching them to build with biology.

No synthetic material or machine we have built to date can match the level of immense complexity and innate responsiveness observed in biological systems.

Building with biological materials requires understanding biology. Fortunately, our curiosity about the natural world has driven tremendous advances in this discipline, generating a growing body of knowledge on the building blocks that make up biological systems. We know how genetic information is coded in DNA and how information coded in genes is translated into the proteins that govern the structure and function of living cells. We have studied the interfaces and interactions that drive the hierarchical assembly of living cells into large functional units, like tissues and organs. We have observed and manipulated the ways in which large organ systems interact with one another in our bodies, both in healthy and diseased states. While we still have much to learn about how so many small parts function in synchrony to generate something as complex and seemingly unpredictable as the human body, we have enough information to start creating a blueprint of how biological systems are built from the bottom up.

A blueprint is, of course, not enough. Building with biology, or *biofabrication*, requires manufacturing technologies that can assemble living materials, an advanced communication and transportation infrastructure that supports a global bio-economy, and ethical and economic policies that govern the safety and sustainability of this technology. Luckily, we live in a world where advances in the tools and infrastructure that enable building with biological materials are converging with our increasing knowledge about biology. We have found new ways to identify and grow different types of living cells. We have created *3D printers* that can manufacture complex shapes using cells as building blocks. We have also developed the ability to make, store, and transport biological materials rapidly around the world. Improved telecommunications, moreover, allow us to engage with our global community instantaneously, enabling rapid information exchange about new technologies and their impact. With this knowledge and these capabilities in hand, we are ideally poised to kick-start the next era of human design and innovation with biofabrication.

Perhaps the most powerful application of biofabrication is in preserving and prolonging human health. Consider, for example, the most advanced prosthetics that can restore mobility to a person who has lost a limb. While effective, they still lack many of the aspects of the original limb that cannot be recaptured using current technologies. Synthetic muscle made of pneumatic actuators can, for example, inflate or contract in response to airflow and generate large forces to produce movement. But, unlike natural muscle, it cannot get stronger when exercised or stay active without an external power source. Replacing biological tissues using synthetic materials alone is unlikely, therefore, to fully restore human mobility and quality of life. A subset of biofabrication known as *tissue*

We live in a world where advances in the tools and infrastructure that enable building with biological materials are converging with our increasing knowledge about biology.

*engineering* has evolved around the idea of replacing damaged or diseased systems in our bodies with engineered tissues built from living cells that are exact replicates of the original tissues. It promises revolutionizing medicine by giving us the opportunity to replace lost limbs with new limbs that have the same form and function as the originals. If this becomes possible, it could radically improve the prognosis for some of the most devastating health challenges we face.

Another emerging medical application of biofabrication is in rapid and safe testing of new experimental therapies. Currently, new therapies such as pharmaceutical drugs are first tested on animals in the lab and then on humans through heavily regulated clinical trials. Finding a new treatment or cure for a disease is an incredibly time-consuming and complex process with a very high failure rate. This is why successfully bringing a new drug from the lab to the clinic costs, on average, between $2 and $3 billion per drug.[1] Researchers are now leveraging biofabrication to develop an alternative approach to this standard by building *organ-on-a-chip* systems. These devices are composed of miniature model tissues that mimic the form and function of larger tissues in the body. New medicines, such as pharmaceutical drugs or gene therapies, can be tested on these mock systems because they closely mimic the structure and interconnected dynamics of organs in the human body. Since organ-on-a-chip devices are very small, they enable cost-effective testing of hundreds of different drugs in the lab prior to conducting a clinical trial to determine safety and efficacy in humans. This means that the therapies that are most likely to be effective in treating diseases can be preselected using these model systems before being tried on real patients. Organs-on-a-chip could thus have a tremendous positive impact on translational medical research, as new therapies can be optimized more safely, quickly, and cheaply than possible with current approaches. This could radically improve the pace of generating new therapies, while concurrently increasing access to health care by reducing development costs.

Biofabrication also has many interesting applications beyond the medical realm, such as in the sustainable production of food. Animal farming to produce meat for human consumption has become an increasing source of contention for many reasons, ranging from concerns about negative environmental impact to debates about the ethics of animal farming. Being able to biofabricate meat could alleviate these concerns by providing a lab-grown alternative to farming animals, while preserving the look, taste, and nutritional value of traditionally sourced meat. While the disciplines of tissue engineering and organ-on-a-chip systems are very focused on replicating the microscale form and function of the human body, to deepen the potential medical relevance of biofabricated tissues, biofabricated meat is focused on very different technical challenges. It requires rapid, cost-effective, and large-scale manufacture of animal tissues with the macroscale mechanical properties (toughness, stretchiness, chewiness), morphology (texture, mouthfeel), and flavor of traditionally produced meat. Matching the appearance, feel, and taste of meat, while simultaneously reducing its manufacturing price point and increasing its global accessibility, could have a tremendous positive impact on next-generation food sourcing.

Perhaps surprisingly, building with biology is also rapidly gaining traction in the field of robotics. Robots, broadly defined, are machines that can autonomously sense, process, and respond to external inputs in real time. A few researchers have explored replacing certain functional components in robots with biological materials, making *biohybrid* machines that leverage the desirable properties of living tissues to perform complex functional tasks. While this technology is still in a very early stage of development, it offers a unique opportunity to explore using biological materials outside of natural contexts. Tissue engineering, organs-on-a-chip, and lab-grown meat all rely on reverse engineering materials and systems that already exist in nature, while biohybrid robots would require forward engineering, or creating new systems entirely from scratch. If this approach is successful, we could generate biohybrid machines that tackle a wider range of pressing technical challenges than synthetic robots can address alone.

The biofabrication tools and applications discussed above primarily focus on building multicellular assemblies of living cells derived from mammals, but this is by no means the only research avenue being explored in this space. Scientists are developing new tools to edit the DNA that resides within individual *mammalian cells* to help treat diseases with a genetic origin. They are also developing RNA-based vaccines to tackle infectious diseases. Cells derived from nonmammalian sources, like bacteria or insects, are being explored for a variety of applications ranging from gut health to global security. Innovators in design and architecture have even proposed that living cells from plants and fungi could serve as sustainable and responsive building materials. These and other applications, while very interesting, are beyond the scope of this book, which is intended to serve as a primer on the main areas of impact that mammalian cell-based biomanufacturing is likely to have on our daily lives. Much of the scientific terminology and the tools that enable these technological advances, however, are generally applicable to a broader definition of biofabrication that applies to building with nucleic acids, proteins, and nonmammalian cells.

Biofabrication is game changing and is moving rapidly, meaning that it raises many economic, environmental, and ethical questions about how it will shape our global landscape. Research and development in this field is very new, and one could argue that advances in building with biology cannot catch up with innovations being made with synthetic materials in terms of ease, cost, and scalability of manufacture. A counterargument can also be made that a front-end investment in developing biofabrication technology, and concurrently developing the manufacturing and transportation infrastructure to support it, could result in drastically better quality of life for current and future generations. Beyond such economic arguments are the environmental concerns associated with building with biology. Manufacturing machines with organic materials that rely solely on sugar and proteins to function sounds like an appealing sustainable alternative to fossil fuel–generated power or to batteries that produce nonbiodegradable waste products. The environmental impact associated with building and operating biomanufacturing facilities, however, as well as with storing and transporting such systems around the world, may outweigh the advantages outlined above. Last, and perhaps most important, building with living cells raises many ethical questions. There is an important distinction between using a cell as a functional building block for a specific device, in a similar fashion to the synthetic materials engineers build with currently, and building autonomous lifeforms capable of conscious decision-making and reproduction. It is important for researchers, policy makers, philosophers, and the general public to reach a consensus on where this distinction lies and to ensure that global guidelines for ethical biofabrication are established, communicated, and uniformly followed.

Biofabricating safe, ethical, and accessible solutions to technical challenges requires active engagement from the global community. To truly understand and govern how building with biology will shape our technology landscape, we must work together to cultivate biofabrication literacy and openly share new information about this discipline. Many lives and communities will be revolutionized by biofabrication, and this book serves as a quick guide to how this technology could improve human health, productivity, and quality of life. By understanding the motivations, current progress, and future directions of this field, you will be better poised to decide how biofabrication will shape the world of tomorrow.

# ENABLING TOOLS AND TECHNIQUES FOR BIOFABRICATION

Building with living cells is no easy feat, though our bodies do this every day. Cells derived from mammals, which are the focus of this book, only function well within a very stringent set of environmental conditions, including precisely regulated temperature, humidity, and pH. These conditions and a whole host of other physical and biochemical factors must be tightly controlled to match the native environment in which such cells have learned to survive and thrive in nature. Thankfully, a large body of research on cell culture, the practice of growing and working with living cells in the lab, has turned this seemingly Herculean feat into a relatively mundane protocol with defined rules and procedures.

Mammalian cell culture must be performed in a sterile, or aseptic, environment to prevent infections that change the behavior of the cells or trigger an early death.[1] It is

Cells derived from mammals only function well within a very stringent set of environmental conditions, including precisely regulated temperature, humidity, and pH.

thus usually performed within a *biosafety hood*, a cabinet outfitted with a high-efficiency particulate air (HEPA) filter and a fan that circulates sterile air within the working area (figure 1). Any object brought inside the hood's working area (including a user's gloved hands) must be sterilized, generally by spraying the outside surface of the object with a solution of 70% ethanol. This extremely

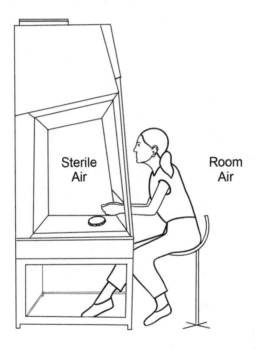

Sterile Air

Room Air

**Figure 1** A scientist handling living cells in a petri dish within the sterile air of a biosafety hood. Art by Radha Raman.

strict regimen must be followed without exceptions to maintain an aseptic environment for the cells. Indeed, after spending years working in a biosafety hood, I started spraying my gloved hands with ethanol every few minutes, just by force of habit, and have seen many others do the same! When not in active use in a biosafety hood, cells are stored inside incubators that maintain the temperature (37°C), humidity (95%), and carbon dioxide concentration (5%) that most closely mimic physiological conditions.

Beyond a sterile, warm, and humid environment, living cells require appropriate nutrients to survive. Cells are usually immersed within *culture media*, which is simply water supplemented with a mixture of glucose, salts, vitamins, and amino acids.[2] Most current formulations of culture media are based on an original recipe first published by Harry Eagle in 1955. Perhaps the most popular modification of Eagle's invention is Dulbecco's Modified Eagle Medium (DMEM), which increases the amount of glucose, vitamins, and amino acids in the original recipe and also adds iron to the formulation. Other modifications, such as Minimum Essential Medium Eagle (α-MEM), and other recipes, such as Roswell Park Memorial Institute's growth medium (RPMI 1640), have also been developed and used in a variety of applications. The culture medium chosen for a specific experiment or application is generally determined by the type of cell being used.

Most media formulations also include a special dye called phenol red. This dye is red in color at the neutral pH of 7, the environment in which most mammalian cells are evolved to function, and turns yellow when the pH drops below 6.8. This drop in pH indicates that the cells are secreting waste materials into the media as a result of metabolic activity. The waste materials turn the media acidic, and the subsequent dye color change from red to yellow serves as an indication that the media should be refreshed. How often the media is refreshed is very dependent on the type and number of cells grown in a culture receptacle. On average, researchers work with protocols that require media to be replenished every day or at least a few times a week, but this frequency can vary widely based on the parameters of their experiments.

An important and somewhat poorly understood component of cell culture media is a material known as serum, generally derived from a *bovine* or other animal source. This serum is extracted from the blood of an animal and contains a wide variety of proteins and growth factors that help cells maintain themselves and grow. While adding serum to cell culture media has many positive effects on the health of the cells, it comes with the drawback of varying in composition between animals, meaning that different batches of serum delivered to labs are not identical to one another.[3] Researchers have tried to combat this by developing serum-free media with formulations that include a combination of nutritional factors. These chemically defined media come with the advantage of having a known composition that does not vary between batches. Unfortunately, these media formulations have yet to be proven as broadly effective for all cell types as media containing serum. Many of the current approaches used to manufacture systems with living cells, therefore, still use culture media containing animal-derived serum.

Living cells cultured in the lab are sourced from a range of tissue types, such as connective tissue, muscle tissue, bone tissue, and blood vessel tissue, to name a few. Cells are not only characterized by the type of tissue from which they are derived but also by the way in which they are derived. Cells used in lab settings tend to fall into three broad categories: *primary cells*, *immortalized cell lines*, or *stem cells*. Primary cells are cells that have been obtained directly from a living animal and thus have many of the important characteristics of the cells in that animal's body. These cells have limited ability to grow and replicate themselves in culture media, however, resulting in finite life spans in the lab.[4] Immortalized cell lines are cells that were originally derived from an animal but have developed a mutation that lets them proliferate and divide indefinitely in culture media, making them very useful tools for biological experimentation. The mutations that make these cells easy to work with, however, also make them a little different from the original source cells, and this must be taken into consideration when drawing biological conclusions based on experiments performed on immortalized cell lines.[5] Stem cells can also grow indefinitely in culture media and have the ability to differentiate into many types of cells inside the body, an ability known as pluripotency. A relatively new advance in stem cell science is the development of *induced pluripotent stem cells* (iPSCs), which provide a way to reprogram mature cells derived from an adult animal into a pluripotent stem cell state.[6] While these cells offer both prolonged culture life spans and relevance to the human body, they can be quite difficult and expensive to grow, maintain, and direct toward a specific cell type. The choice of which category of cell to use for a specific application, be it in medicine, food, or robotics, is determined by the functional requirements of the application, as is discussed in the following chapters.

For many years, scientists have grown cells inside plastic or glass dishes coated with molecules that encourage cell adhesion. This technique is called 2D culture, because the cells grow in a single layer on the surface of the dish. Patterning molecules in different places or shapes on a dish's surface enables precisely controlling where cells attach and how they grow. This type of 2D patterning has been used to culture different types of cells in the same dish, control how they spatially interact with one another, and study the resultant behavior.

While 2D culture methods have generated significant biological insights, they do not fully mimic the native environment in which mammalian cells function. The cells in our body, for example, function in complex 3D structures, and, in recent years, many researchers have transitioned to culturing cells in 3D platforms that more realistically mimic their native environment.[7] In the multicellular tissues found in our bodies, cells are surrounded by a structure known as the *extracellular matrix* that provides both physical and biochemical support. Mimicking the form and function of the proteins and enzymes that form the extracellular matrix in a lab setting has fueled research in *hydrogels*, polymer matrices that are soft, wet, and resemble some of the important physical properties of native tissue.[8] Most 3D culture methods involve embedding living cells within hydrogels to form multicellular structures.

Hydrogels derived from both natural and synthetic sources have been used for this purpose. Natural sources include materials like *collagen*, a protein found both in native extracellular matrix and other connective tissues, and *fibrin*, a mixture of the protein fibrinogen and the enzyme thrombin that form clots to prevent bleeding after an injury. Some nonmammalian natural hydrogel sources, such as alginate (derived from algae) and chitosan (derived from crustacean shells), have also been successfully used as support environments for mammalian cell culture.[9] There is also a very broad range of synthetic

In recent years, many researchers have transitioned to culturing cells in 3D platforms that more realistically mimic their native environment.

jellylike polymers that have been adapted for cell culture, such as polyethylene glycol (PEG), polyacrylamide (PAM), and polyvinyl alcohol (PVA).[10] Naturally derived hydrogels come with the advantage of having many of the biochemical factors that preserve and promote cell growth, as well as being biodegradable. Synthetic hydrogels, by contrast, have the advantage of possessing highly tunable biological and mechanical properties. These hydrogels can be chemically modified to contain different types and quantities of biological factors or mechanically modified to be as stiff or as stretchy as desired. As with the cell source and culture media chosen for any given experiment, the type of hydrogel chosen for a particular biofabrication application is highly dependent on the functional characteristics that are most desired.

Forming cells embedded within hydrogels into complex 3D shapes is a requirement of many of the applications we will discuss in this book, motivating a discussion of some of the tools that enable manufacturing such multicellular systems. These technologies were all initially developed for building with other materials, such as metals and plastics, but have since been adapted for use with *biocompatible* materials (like synthetic hydrogels that are not harmful to living cells) and biological materials (like living cells and naturally derived hydrogels). The predominant biomanufacturing techniques used today are based on injection molding, microfluidic platforms, and 3D printing.[11]

If you have spent any amount of time experimenting in a kitchen, you probably have personal experience with injection molding. You may know that gelatin-based dessert, like Jell-O, is formed by pouring boiling water mixed with gelatin (and sugar and dye and other flavors) into a mold and then letting it set into a solid 3D shape before removing it from the mold. This is how living cells are embedded in hydrogels with injection molding, except that the hydrogels are converted from liquid to solid states in less harsh environmental conditions that keep cells healthy. After all, boiling living cells, as one does to mold a gelatin dessert, would kill them! Injection molding for biofabrication thus involves using hydrogels that turn from liquid to solid in conditions more easily tolerated by cells. Gelatin is usually a solid below 15°C and starts melting at 37°C (body temperature), so raising the temperature to the level at which water boils (100°C) is not actually required.[12] Some polymers even go from liquid to solid when they are heated, instead of cooled. For example, Matrigel, a commonly used naturally derived hydrogel that mimics the extracellular matrix in the body, turns from a liquid at 4°C to a solid at 37°C (body temperature).[13] Gelation methods based on chemical mechanisms, rather than temperature changes, have also proved very useful in injection molding. The aforementioned fibrin hydrogels, for example, can be formed by mixing together liquid fibrinogen and liquid thrombin.[14] Injecting the liquid mixture into a mold results in a solid 3D fibrin hydrogel formed within the mold. While injection molding offers a very simple and robust mechanism for manufacturing cell-laden gels at the millimeter or centimeter scale, making biofabricated systems that are much smaller or much larger require different manufacturing approaches.

Microfluidic platforms offer an attractive and elegant approach for manufacturing miniature biofabricated constructs. Microfluidic platforms were initially developed for manipulating very small amounts of liquid, with volumes on the order of microliters (a microliter is a millionth of a liter). There are many ways to manufacture such a device, but most *microfluidic devices* are made by casting a silicone polymer on top of a special mold with tiny embossed lines. The cast silicone, as a result, contains tiny etched lines that form fluidic channels when bonded with a piece of glass (figure 2). Advances in microfluidic design and fabrication techniques have resulted in several complex passive and active components that can be incorporated with these devices. These include elements, such as interfacing channels, valves, mixers, and pumps, that enable complex fluid manipulations with extremely precise volume control.[15] In the context of biofabrication, this technology enables us to mix together different types of cells and the liquid precursors of hydrogels inside microfluidic channels, forming microscopic 3D cell-gel constructs.[16] Nutrient transport to and waste transport from these cell-gel constructs can

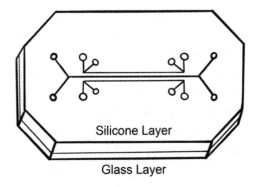

**Figure 2** A microfluidic device formed by bonding a layer of silicone containing microscopic etched lines with a thin layer of glass. The space generated by the etched lines between the silicone and the glass results in a microscopic fluidic channel. Art by Radha Raman.

be controlled with extraordinary precision by flowing cell culture media through the microfluidic channels. Different channels can even be used to flow different types of media, enabling the culture of many different types of cells within a single device. This proves particularly helpful in the design, manufacture, and manipulation of organ-on-a-chip platforms, as is discussed in chapter 4.

Perhaps the most sophisticated biomanufacturing technology, and the one that best enables building large multicellular tissues with complex internal microscale architecture, is 3D printing. The term 3D printing is applied to a broad array of technologies that rely on different mechanisms to construct complex 3D structures.[17]

The feature that unites most of these technologies is the basic principle of building a 3D structure by stacking 2D layers on top of one another. Revisiting the concept of kitchen experimentation, let us think about how we pipe icing onto cakes. A bag with a nozzle is filled with icing, and squeezing the bag results in icing flowing out of the nozzle in a controlled manner. You can use this relatively simple technique to create a complex 2D design that turns from liquid to solid as the sugary icing hardens. If you pipe another layer of fresh icing onto a layer of hardened icing and let it harden, and then pipe yet another layer of fresh icing onto the previous two hardened layers, you can build a solid 3D structure made out of icing. This is the fundamental principle of extrusion-based 3D printing (figure 3).

Imagine if, instead of icing, the nozzle was depositing a liquid hydrogel precursor that could turn solid using any of the temperature- or chemical-based mechanisms discussed above. Turning each deposited liquid layer into a solid before depositing the next layer enables building a complex 3D structure, layer by layer, from the bottom up. Early experiments in using this approach with cells and biocompatible hydrogels, conducted in the lab of Gabor Forgacs around 2009, showed that this approach could be used to manufacture small hollow tubes that resembled blood vessels.[18] The diameter and length of these tubes, as well as the materials used to make them, could be readily modified as desired for a target application. Since the

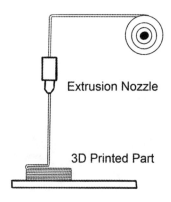

**Figure 3** An extrusion-based 3D printer dispenses a liquid solution through a nozzle, which solidifies when deposited on a surface. This technique can be used to build 3D structures, layer by layer, from the bottom up. Art by Radha Raman.

hydrogels used for biofabrication are very soft by design, so that they match the mechanical properties of native tissue, they can often collapse on themselves or deform as they are printed. This becomes a significant concern, especially as printed structures become larger and require the assembly of several layers on top of one another. To combat this, Adam Feinberg's lab started extruding hydrogels into a support bath in 2015 and named this printing technique FRESH: freeform reversible embedding of suspended hydrogels (figure 4).[19] With FRESH printing, soft hydrogels are extruded from a nozzle into a support bath that preserves them from deforming. This enables building larger structures out of hydrogels without them

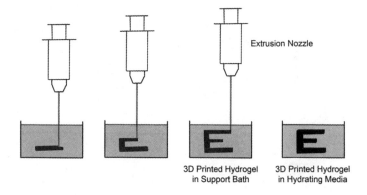

Extrusion Nozzle

3D Printed Hydrogel
in Support Bath

3D Printed Hydrogel
in Hydrating Media

**Figure 4** A FRESH printer generates 3D structures from soft materials, like hydrogels, by printing consecutive 2D layers in a support bath that prevents each layer from deforming during fabrication. Upon release from the support bath, the 3D printed part can be stored in a liquid medium that keeps the structure hydrated. Art by Radha Raman.

bending, buckling, or changing shape during fabrication. After a structure is fully printed, the support bath, which is made of a gelatin slurry, is melted by heating it to 37°C. The printed structure is thus gently released and can be stored in a liquid medium to keep it hydrated. If the printed structure contains cells, it is immersed in a warm cell culture medium that preserves and promotes cell health.

Nutrients from the cell culture media, as well as oxygen from the surrounding environment, are transported to cells embedded within 3D hydrogels by passive diffusion. The distance through which these molecules can diffuse into 3D cell-gel structures is on the order of a quarter of a millimeter, meaning that cells in the center of a cylinder larger than half a millimeter in diameter would not receive the nutrients they need to survive.[20] As a result, an ongoing concern with printing large 3D cell-gel structures is finding ways to regulate oxygen and nutrient transport.

We transport oxygen and nutrients throughout our bodies using the vascular system, the network of blood vessels that are present in nearly all our tissues. These blood vessels range in size from a few micrometers to several millimeters in diameter. Manufacturing a vascular network within a 3D cellular construct requires printing with several different types of cells and being able to manufacture vessels with very small feature sizes. Many researchers are pursuing different methods for printing tissues with embedded blood vessels. A promising technique from Jennifer Lewis's lab in 2019 demonstrated the ability to manufacture large tissues containing vascular networks using a modified version of FRESH printing.[21] This technique, termed *SWIFT*, sacrificial writing into functional tissue, also involves extruding a material into a support bath. In this case, however, the support bath is not a sacrificial material that is removed postfabrication but rather is composed of cells mixed with the naturally derived hydrogels, collagen and Matrigel. The printer nozzle is inserted into this cell-gel tissue-like structure and extrudes a sacrificial ink that, when removed, leaves behind hollow channels in the tissue. Cell culture media can be perfused through these channels, thereby distributing nutrients and oxygen throughout the thickness of the tissue.

Continuing advances in these and other extrusion-based 3D printing approaches will likely enable manufacturing large (centimeter-scale) vascularized living tissues in the coming years. There are certain biomanufacturing applications, however, that require printing with multiple types of cells in different layers or printing microscale features within large tissues. These applications may be best served by a different type of 3D printer. One of the most common types of 3D printing techniques is termed *stereolithography*. Stereolithography relies on polymers that turn from liquid to solid when exposed to light at certain wavelengths.[22] This technique, patented in 1986 by Chuck Hull, also relies on a layer-by-layer fabrication approach. Stereolithographic printers are generally composed of a vat filled with a light-sensitive liquid, with a printing stage immersed in the vat. A laser traces a 2D design onto the liquid, and every region touched by the laser turns from liquid to solid. The stage moves deeper into the vat, bringing a fresh layer of liquid on top of the solidified layer. The laser traces the liquid surface again, solidifying another 2D layer. Sequential patterning and solidification of 2D layers on top of one another results in the fabrication of a 3D structure, layer by layer, from the bottom up

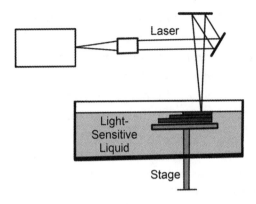

**Figure 5**  A stereolithographic 3D printer traces a laser across a vat of light-sensitive liquid to turn traced regions solid. As the stage upon which the 3D printed part is being built moves deeper into the vat, it brings a fresh layer of light-sensitive liquid on top of a previously solidified layer, allowing sequential construction of the part from the bottom up. Art by Radha Raman.

(figure 5). One of the primary advantages of stereolithography is that the printing material can be changed from layer to layer, with a simple wash step between each printed layer. In the context of biofabrication, this means that different cells can be encapsulated in different layers, enabling manufacturing complex 3D structures.

Preliminary experiments by the labs of Thomas Boland in 2004 and Ryan Wicker in 2006 showed that stereolithography could be adapted to printing with cells seeded within a synthetic hydrogel, PEG.[23] These hydrogels, when mixed with a chemical known as Irgacure

2959 (chemical name: 2-hydroxy-1-[4-(2-hydroxyethoxy) phenyl]-2-methyl-1-propanone), become light reactive and can turn from liquid to solid upon exposure to ultraviolet light (365-nanometer wavelength). Kristi Anseth's lab, in 2009, developed an alternative to Irgacure 2959 called lithium acylphosphinate (LAP) that turned PEG-based hydrogels from liquid to solid much faster and worked with visible blue light at 405 nanometers.[24] This was an important advance because living cells can be quite sensitive to prolonged exposure to light at ultraviolet wavelengths. This is why we are all encouraged to wear protective clothing, sunscreen, and sunglasses to block ultraviolet rays from the sun.[25] By enabling biofabrication with blue light in the visible spectrum, therefore, LAP could potentially increase the biocompatibility of stereolithographic approaches by decreasing cell exposure to ultraviolet light.

Since lasers are used to trace across a light-sensitive polymer in traditional stereolithography, the smallest feature that can be manufactured is the diameter of the laser beam, usually on the order of a few hundred micrometers.[26] Printing features that are similar in size to living mammalian cells, which are generally less than 100 micrometers in diameter, thus requires a different approach. One promising approach for high-resolution 3D printing is called projection stereolithography. Projection stereolithographic systems leverage the same technology that powers the digital projectors you may have encountered in classrooms, offices, or movie theaters. Projectors are powered by small components known as digital micromirror devices (DMDs). DMDs are made of an array of microscale metal mirrors that can be individually rotated to be turned on or off. On is represented by the color white, and off is represented by the color black, so individually rotating each mirror in an array can be used to produce a 2D grayscale image. The digital projectors in classrooms and movie theaters magnify these 2D grayscale images by projecting them onto a wall or screen, generally adding color. Projection-based 3D printers, however, demagnify these 2D images (i.e., make them smaller) by focusing them through a lens and projecting the miniaturized image onto a light-sensitive polymer. White areas polymerize from liquid to solid, and black areas remain liquid. An entire 2D design can thus be printed in a single exposure. By enabling the manufacture of complex 2D patterns with high resolution in a single step, this technique makes it easy and fast to 3D print structures with intricate internal architectures.

Shaochen Chen's lab showed, in 2012, that projection stereolithographic printers could be used to manufacture complex 3D structures from biocompatible materials that could serve as scaffolds for growing living cells.[27] My own research in Rashid Bashir's lab showed, a few years later, that such printers could also be modified to directly print cells embedded within hydrogels.[28] Using this projector-based technique instead of a laser-based approach enabled printing features within hydrogels that were less than five micrometers in size, the diameter of the smallest blood vessels inside our bodies. Interestingly, our experiments showed us that building a part upside down enabled quickly changing the printing material between each layer without an intermediate wash step. This meant that different cells, with different forms and functions, could be printed within each 2D layer at high resolution, resulting in an easier way to build complex multicellular 3D structures.

As is hopefully evident from this quick overview of the various manufacturing technologies used for biofabrication, each tool has its advantages and disadvantages.[29] Injection molding can be used to rapidly fabricate meso-scale 3D structures but cannot be readily adapted for multimaterial printing or creating vascularized tissues. Microfluidic devices can be used to manufacture tissues with precise microscale architectures but cannot readily be scaled up to manufacture millimeter- or centimeter-scale tissues. Extrusion-based 3D printing can be used to manufacture larger tissues, even containing perfusable vascular systems, but cannot be easily tuned to use many different materials in a single construct or to incorporate microscale features. Stereolithographic 3D printers enable multimaterial manufacturing and microscale patterning but require exposing cells to light-sensitizing chemicals that reduce biocompatibility and often cannot pattern cells at very high density compared to other approaches.

Furthermore, all the manufacturing approaches presented above must operate within very tight tolerances of temperature, pH, and sterility, as discussed earlier in this chapter, to be used for biofabrication. This limitation increases the difficulty of using very sensitive cell types or building large tissues that take a long time to manufacture. As with most other technologies in our world, there is always a tradeoff in positives and negatives for each biofabrication tool presented in this chapter. Different manufacturing approaches are used for different target applications, with varying degrees of success, as further explored in chapters 3 through 7. It is likely that many of these approaches will be developed into more robust, facile, and affordable tools in the coming years and may start impacting our daily lives in the near future.

# TISSUE ENGINEERING

Having the tools to build complex 3D structures with living cells opens up a world of possibilities. Perhaps the most immediately compelling possibility is replacing diseased or damaged tissue in the body with engineered tissue that replicates the form and function of healthy tissue. This is the simple yet powerful core mission of the field of tissue engineering, also referred to as regenerative medicine. Using biofabrication tools to reverse engineer our bodily systems has driven many of the seminal advances in this field.[1]

The tissues inside our bodies have varying levels of ability to regenerate after disease or damage. Cartilage tissue that lines and cushions our joints, for example, has limited ability to self-heal since it is avascular, or lacking in blood vessels. Cardiac muscle tissue that powers the beating of our hearts is also unable to regenerate, as adult human cardiac muscle cells cannot grow and replicate. This is why diseases like osteoarthritis, in which worn-down cartilage results in joint pain and stiffness, and heart attacks, where loss in blood supply results in the death of cardiac muscle tissue, have such severe and lasting negative consequences.[2] Some tissues, like skin, skeletal muscle, and bone, have a greater ability to regenerate after damage. We all know that our bodies can readily recover from small scratches, muscle tears, and bone fractures, returning these tissues to their original form and function. Much larger injuries to skin, muscle, and bone, such as extensive burn damage or the loss of a limb, however, generally result in an irrecoverable loss of health, mobility, and quality of life.[3] Certain types of tissue inside our bodies can recover from very large amounts of damage. The liver, for example, possesses the remarkable ability to recover its full size and function from only a quarter of the volume of the original tissue. If the liver is damaged by toxic agents as a result of alcohol or drug abuse, however, it develops scar tissue (severe scarring is known as cirrhosis) that can prevent natural regeneration.[4] For nearly every type of tissue in our bodies, therefore, damage caused by disease or injury can limit one's ability to enjoy a healthy life. The field of tissue engineering has thus evolved around developing replacements for all the tissue types discussed above, as well as many others.

For nearly every type of tissue in our bodies, damage caused by disease or injury can limit one's ability to enjoy a healthy life.

Many early tissue engineering experiments focused on reverse engineering cartilage. While the lack of an internal blood supply limits the ability of cartilage to regenerate in our bodies, it also makes it a bit easier to reverse engineer, as there is no need to microfabricate the complex architecture of blood vessels that permeate most other tissues. Perhaps the most well-known tissue engineering experiment is the Vacanti mouse, developed by the lab of Charles Vacanti in 1997.[5] The Vacanti lab placed living cartilage cells, called chondrocytes, in a synthetic polymer scaffolding formed in the shape of a child's ear. The engineered tissue was then implanted under the skin of a mouse's back, and the chondrocytes grew and proliferated within the scaffold, giving the appearance of a mouse growing a human ear on its back. This seminal experiment—and the highly publicized images of the Vacanti mouse—helped kick-start the field of tissue engineering.

The Vacanti experiment's approach of first manufacturing a scaffold and then seeding it with living cells was quite popular in the early days of tissue engineering and was tested for regenerating many different types of tissues. In recent years, however, it has become more common for scientists to pattern cells and scaffolds simultaneously, as with the cell-gel constructs discussed in chapter 2. This shift in the field was triggered by the design and development of new biocompatible polymers, several of which are discussed in this chapter, that could provide cells with long-term mechanical and biochemical support.

Lawrence Bonassar's lab, for example, has utilized simultaneous cell-gel patterning for applications in cartilage tissue engineering. The lab has used injection molding to form alginate hydrogels with embedded chondrocytes in a variety of geometries. By using an injection mold designed from a 3D scan of a real person, they have demonstrated that chondrocytes derived from sheep can be grown in the shape of a human ear.[6] When these cell-gel constructs were implanted in the sheep from which the cells were harvested, they were shown to be well tolerated by the animal's body and to encourage the formation of new cartilage while retaining the overall 3D structure of the human ear. The lab has used this method to engineer other types of cartilage, such as the tissue that forms intervertebral discs (the tissue between adjacent vertebrae in the spine) and knee menisci (the tissue that cushions the joint between your femur and tibia).[7]

Using scans from real people to develop injection molds enables building tissues that are patient specific, or exactly the right size and shape for each individual's needs. However, injection molding produces tissues that are homogeneous, with similar composition and properties throughout the tissue volume. Researchers in Bonassar's lab, including myself in my undergraduate days, hypothesized that exerting mechanical forces on the molded tissues in a manner similar to how they are loaded inside the body could help generate a more complex internal structure that mimicked anatomical organization. We tested this by loading engineered knee meniscus cartilage into a custom-designed *bioreactor*, or tissue support system, that applied forces to the tissue, mimicking the forces imposed on the meniscus by the knee.[8] Our team observed that loading engineered menisci with this artificial knee resulted in the embedded cells and proteins self-organizing in a way that better recreated the mechanical properties of native tissue. This is one of many experiments that showed us and others in the field that biofabricated constructs are not static. Biological materials, unlike synthetic materials, rely on dynamic cues from their environment to inform their form and function and can adapt to their surroundings as needed.

Dynamic cues from the environment can also take the form of biochemical stimuli. In 2006, Jennifer Elisseeff's lab developed a technique for using a biochemical stimulus to tissue engineer cartilage from a stem cell source.[9] The team embedded and grew stem cells inside a PEG-based hydrogel manufactured using a modified stereolithographic approach. Exposure to ultraviolet light turned the liquid PEG, which had been premixed with stem cells, into a solid gel that surrounded and supported the embedded cells. The lab grew the engineered tissue in a cell culture

Biofabricated constructs are not static. Unlike synthetic materials, [they] rely on dynamic cues from their environment to inform their form and function and can adapt to their surroundings.

media that contained biochemical factors known to drive stem cells to mature into cartilage cells and showed that the cells started depositing extracellular matrix components that were characteristic of newly formed cartilage. It is likely that a combination of both biochemical cues, as illustrated in this experiment by Elisseeff's lab, and mechanical cues, as tested in the experiment by Bonassar's lab, will be used in the future to generate tissue replacements that replicate both the structure and strength of the cartilage in our bodies.

Moving past cartilage toward other tissues, such as cardiac tissue, showcases other advances in this burgeoning field. Native cardiac muscle can be thought of as a sheet of individual cells, where adjacent cells are electrically coupled to each other via connections known as gap junctions. This electric connection ensures that all the cells contract at the same time, which in turn drives the synchronous beating of our hearts. For this reason, many approaches targeted at regenerating heart muscle in the lab revolve around the idea of cardiac tissue sheets, or patches, that can be placed on top of scarred regions of a damaged heart. In order to function appropriately after implantation, the tissue sheets must form electrical connections with the surrounding tissues. They must also allow for blood vessels to grow into the patch, so that the vascular network in the native tissue connects through the engineered tissue, ensuring oxygen and nutrient transport through the implant.[10]

Thomas Eschenhagen's lab made early advances toward tissue engineering cardiac muscle in 2002 by embedding primary heart cells from rats (*cardiomyocytes*) in a natural hydrogel matrix composed of collagen and Matrigel.[11] The team noted that the cells started contracting, or beating, a day after the tissue was formed and that the contractions started occurring in synchrony about two days after tissue formation. Around the same time, the lab of Gordana Vunjak-Novakovic engineered tissue constructs using the same type of primary cells but a different type of extracellular matrix, a scaffold composed of polyglycolic acid.[12] Her team showed that actively perfusing the engineered tissues with culture medium improved the transport of oxygen and other nutrients and metabolites to the embedded cells. This enabled the construction of thicker cardiac tissues that could prove easier to handle during an implantation procedure. In later years, they showed that culturing the cells in a highly porous matrix could also enhance the flow of the surrounding culture medium into the tissue.[13]

These advances showed that we could engineer cardiac muscle in the lab and keep it alive and functional, but there was still much left to learn about understanding and controlling the electrical coordination between cardiomyocytes in engineered tissues. Milica Radisic and colleagues recognized in 2004 that, since cardiac muscle is electrically active in the native heart, it would be important to integrate electrical signals with the development and maturation of heart tissues in a lab setting.[14] To test this, they biofabricated cardiac muscle in the lab and applied electrical signals to the tissue that mimicked physiological conditions. This electrical stimulation increased the structural organization of the cells in the engineered tissue, as well as the strength of the contractions they produced. In 2013, Ali Khademhosseini's lab built on this idea by incorporating carbon nanotubes (CNTs) into the hydrogels with which they were constructing engineered cardiac tissues.[15] The CNTs were electrically conductive and formed nanoscale fibrous networks within the hydrogels that helped electrically couple the embedded cardiac cells to each other. This sort of hybrid approach to biofabrication, which leverages both biological and synthetic materials to fabricate complex engineered tissues, has increased in popularity in recent years.

While the approaches outlined above used primary cell lines from animals to construct engineered heart tissues, there is also a significant body of research on deriving cardiac muscle cells from human iPSCs, as this will be critical to translating tissue engineering research to the clinic. This is because using a patient's own cells to treat them will reduce the likelihood that the patient has a negative immune response to an engineered tissue when it is implanted.[16] iPSC approaches rely on harvesting cells from a patient via a skin biopsy or other minimally invasive technique, reprogramming them to form stem cells, and then maturing the stem cells into cardiomyocytes. While stem cell–based approaches are being refined, the experiments performed on primary cells derived from other mammals can still generate tremendous insight into the best way to engineer replacement tissues for cardiac muscle, as well as for other tissue types, such as the liver.

Engineering liver tissue comes with many of the same design and manufacturing considerations that are relevant for other tissue types but also has some unique challenges that must be overcome. Primary liver cells, or *hepatocytes*, have been shown to lose their liver-like functions when they are maintained in the standard cell culture conditions outlined in chapter 2. Several early efforts in this field, through the 1980s and 1990s, attempted different approaches to mitigate this loss in function.[17] These efforts included growing the cells in 3D scaffold environments, culturing them alongside different cell types, placing the engineered tissues in moving flasks to enhance media flow and resultant nutrient transport, and other approaches that had driven advances in engineering different types of tissues.

A significant step forward in this field was demonstrated by Linda Griffith's lab in 2001. The team showed that a custom bioreactor that controlled flow of culture media across the tissues could help generate a better engineered liver.[18] Hepatocytes from rats were placed within tiny wells and started forming microscale liver tissues

Using a patient's own cells to treat them will reduce the likelihood that the patient has a negative immune response to an engineered tissue when it is implanted.

within the wells in about a day. A porous filter formed the bottom of the wells, and a bioreactor flowed cell culture media across the top of the wells. The media perfused through the tissues and the underlying porous filters before flowing out of the bioreactor. This active fluid flow through the tissue at a defined and uniform rate improved the robustness and liver-like function of the engineered tissues by enhancing nutrient transport and exerting forces on the cells that resembled what they experience in the body. In 2014, I helped the Griffith lab change the material of their wells from a rigid silicon to a PEG-based hydrogel, which was softer and more closely resembled the compliant nature of native tissue.[19] This combination of a soft scaffolding material, coupled with the bioreactor-driven controlled flow of media through the liver tissue, enhanced the levels of albumin protein produced by the liver cells. As the production of albumin is an important part of normal hepatocyte function, this served as a promising indication that this technique could be used to induce and maintain native liver-like function in engineered liver tissues.

Approaches that embed hepatocytes directly within hydrogels have also been explored as a potential strategy for developing engineered liver tissues. Smadar Cohen's lab embedded primary rat hepatocytes within alginate hydrogels in 2003 and discovered that the density at which cells were seeded in the gels had a significant effect on their viability.[20] If the density was too low, the cells could not aggregate to form multicellular clusters and died within a day. Increasing the density of cells in the hydrogel enabled the formation of multicellular spheroids and helped the cells establish physical and biochemical contact with their neighbors, thereby improving viability and albumin secretion levels. Yoshiyuki Nakajima and colleagues relied on this idea of improving cell-to-cell contact to manufacture large-scale engineered liver tissue in 2007.[21] Primary hepatocytes from mice were cultured on dishes coated with a temperature-responsive hydrogel, poly (N-isopropylacrylamide) (PIPAAm), and grew and proliferated to form a uniform sheet of cells. By lowering the temperature of the dish, the team could detach the hydrogel (and attached tissue sheet) from the dish's surface while maintaining the integrity of the 2D cell layer. They showed that many layers of liver tissue sheets could be stacked on top of one another to form a thick 3D liver tissue. When the engineered liver tissues were implanted in mice, they remained viable and produced albumin for more than 200 days. Moreover, surgically removing two-thirds of the engineered liver after implantation resulted in the tissue growing back to counteract the effect of the induced damage. This demonstrated that the engineered liver maintained the impressive regenerative functionalities of native liver, validating many of the hopes and promises of tissue engineering approaches

Notably, this type of autonomous regeneration after damage is not unique to liver tissue. Bone has a tremendous ability to regenerate and is in fact constantly remodeling itself throughout our lives in response to small damages and changing environmental loads. In other words, our bones get stronger when we exercise and can recover from fractures and breaks. Very large damages caused by severe injuries or surgeries such as bone tumor resection, however, cannot be simply regenerated without external medical intervention.[22] As a result, several approaches for generating functional replacements for diseased or damaged bone in the body have been explored since the 1980s.

As with most other engineered tissues, engineered bone requires a cell source that can grow into bone, an external physical and biochemical support system that mimics the native extracellular environment, and vascularization. Bone tissue engineering comes with the additional challenge that the implanted tissue must match the physical properties of the surrounding bone. Bone is much stronger and stiffer than most other tissues in our body, which are generally quite soft and jellylike.[23] Since cells can sense their mechanical environment, bone cells require an environment with mechanical properties similar to native bone in order to grow, proliferate, and maintain healthy function.[24] For this reason, though hydrogels have been used in some bone tissue-engineering studies, many studies combine softer hydrogels such as collagen and gelatin with stiffer materials, such as ceramic particles or polymer fibers, to form composite scaffold materials with more bone-like characteristics.

Many different cell sources have been applied to bone tissue engineering, with human mesenchymal stem cells being perhaps the most popular. Mesenchymal stem cells can differentiate into a wide variety of tissues, such as bone, cartilage, muscle, and fat, and are generally derived from bone marrow. A collaborative effort between the labs of Gordana Vunjak-Novakovic and Robert Langer in 2004 investigated culturing mesenchymal stem cells (derived from donated human bone marrow) in a collagen and silk fiber-based matrix.[25] They discovered that, while culturing these cells in 2D films and 3D scaffolds both resulted in the cells differentiating into a bone-like state, the cells deposited much more calcium when grown in 3D scaffolds and subjected to active culture media flow within a bioreactor. Since calcium deposition resembles the activity of native bone, this approach was deemed advantageous. The researchers also showed that the addition of silk, which has robust mechanical properties, to the scaffolds helped slow down the rate of scaffold degradation over time, as compared to scaffolds made of collagen alone. Tissue constructs formed purely using cells in a collagen matrix retained only 25% of the DNA in the original construct after a month of culture in the lab. By contrast, collagen–silk constructs retained more than 80% of their initial DNA content after the same period of time, indicating that they may be more useful in applications that required engineered tissue longevity.

These early studies in enhancing media flow through tissue constructs raised a lot of questions about how mechanical stimulation affects bone cells in culture. The lab of Lutz Claes conducted mechanical stimulation studies on cells that were precursors to *osteoblasts*, the cells responsible for bone formation in our bodies, seeded within a collagen gel.[26] The team discovered that cyclically loading the engineered bone tissue—for example, by regularly stretching the tissue—increased the growth and differentiation of the precursor cells into osteoblasts. Others have looked beyond direct mechanical stimulation to the application of other forces, such as electromagnetic fields. Barbara Boyan's lab has studied cultures of human mesenchymal stem cells grown on discs made of calcium phosphate and provided them with a biochemical stimulus (bone morphogenetic protein-2) to induce their maturation into osteoblasts.[27] The team subjected the cells to pulsed electromagnetic fields and showed that this enhanced the ability of the stem cells to mature into osteoblasts. These and other types of dynamic stimuli showcase the truly responsive nature of living cells and remind us that engineering functional replacement tissues requires a deep understanding of the many types of mechanical and biochemical cues that guide tissue formation, function, and regeneration in our bodies.

Improving our understanding of how to engineer different types of tissues must be supplemented by developing robust strategies for vascularizing engineered tissues. Integrating blood vessel networks with engineered tissues is critical to ensuring oxygen and nutrient transport can occur in bigger constructs, which will be required for replacing large damages in the body. Several approaches for integrating perfusable blood vessel–like channels within engineered tissues have been explored, such as the SWIFT printing technique described in chapter 2. A different technique, developed by the labs of Sangeeta Bhatia and Christopher Chen in 2012, utilized a sugar-based carbohydrate glass as a sacrificial template for generating channels inside 3D tissues.[28] They showed that encapsulating a 3D sugar lattice in a cell-gel construct and then dissolving that lattice (by immersing the construct in cell culture medium) left behind open channels within the engineered tissue. These channels could then be lined with *endothelial cells*, the cells that line the interior surface of blood vessels in our bodies, to generate engineered vasculature that sustained the flow of human blood through the tissue.

Another approach to engineering vasculature is to generate fluidic microchannels within tissues after they

Engineering functional replacement tissues requires a deep understanding of the many types of mechanical and biochemical cues that guide tissue formation, function, and regeneration in our bodies.

have been formed rather than during fabrication. Cole DeForest's lab has taken a step toward this goal by developing synthetic hydrogels that can be degraded on demand using different external stimuli. For example, by building on work from Kristi Anseth's lab, they have shown that hydrogels that are modified to include a light-sensitive chemical can degrade in response to light exposure.[29] In 2017, these researchers showed that they could use a laser to selectively degrade portions of a cell-gel construct manufactured using their hydrogels.[30] A light-sensitive hydrogel encapsulating bone marrow–derived cells was degraded with a laser in precise locations, leaving behind a network of channels throughout the 3D construct. These channels were then lined with endothelial cells to form perfusable vasculature, as with the approach outlined above, and shown to be fluidically functional.

In applications where 2D engineered tissue sheets are more desirable than large 3D constructs, directly incorporating vasculature into the engineered tissue during the fabrication process may not be necessary. My research in the lab of Rashid Bashir showed that *fibroblast* cells embedded within hydrogels could be induced to secrete biochemical factors that encourage the formation of new vasculature in surrounding tissues.[31] Fibroblasts are the most common cells in connective tissue and are responsible for synthesizing the extracellular matrix. When fibroblasts are exposed to the chemical tetradecanoylphorbol 13-acetate (TPA), they secrete vascular endothelial growth factor (VEGF), a protein that stimulates the formation of blood vessels. Encapsulating fibroblasts within hydrogels and stimulating them with TPA thus results in a biofabricated tissue that produces VEGF on demand. When we placed these VEGF-secreting sheets on the vascular membrane of a chicken egg, we saw that the number and diameter of blood vessels formed on the membrane increased in the areas surrounding the engineered tissue. In 2017, Hyunjoon Kong's lab showed that similar biofabricated sheets that secreted VEGF could be placed on the surface of a mouse's heart to help recover cardiac function after damage.[32]

These and other studies have shown that, for 2D tissue sheets implanted in the body, promoting the growth of the surrounding tissue's vascular network into the engineered tissue is enough to sustain nutrient transport. This is not the case for thicker 3D engineered tissues, where a vascular network must often be embedded within the tissue prior to implantation to prevent cell starvation and death in the center of the construct. Strategies to connect the engineered vascular network within these 3D constructs with that of the surrounding tissue when implanted in the body are the subject of ongoing exploration.[33] The important thing to remember with all tissue engineering and vascularization strategies is that the protocol employed will be highly dependent on the type of tissue and size of the construct required for a given application.

The examples of tissue engineering outlined in this chapter represent only a small fraction of the many ongoing research efforts for engineering replacements for diseased or damaged cartilage, cardiac muscle, liver, bone, and blood vessels. Moreover, they do not touch on the extensive work being done on engineering other types of tissues, such as the tissues that make up our skin, brains, and kidneys.[34] They do, however, illustrate the multifaceted and dynamic considerations that come into play when the complex tissues inside our bodies must be reverse engineered and built from scratch in a lab setting. Building these in a cost-effective manner at scale will also come with further challenges as is described in chapter 7.

The most daunting nontechnical challenge for tissue engineering will be obtaining approval from the Food and Drug Administration (FDA) in the United States and other regulatory agencies worldwide for implanting engineered tissues inside human bodies. While there are many existing regulatory pathways for testing and approving pharmaceutical drugs and implantable devices, engineered tissues are a whole new class of medicines, and we are still learning how to assess their safety and efficacy. Recent cell therapy clinical trials, in which engineered cells are injected in a patient to generate a positive therapeutic response, have met with significant success and received

FDA approval. These clinical trials are setting the stage for future trials in which engineered cell-gel constructs are regularly implanted in the body to replace diseased or damaged tissue. While clinical trials are inherently unpredictable, the long-term trend for integrating engineered cells and tissues with the human body is positive, and I am hopeful that we will see many life-saving tissue engineering approaches translate to human use in the coming years.

Tissue engineering is certainly no small feat, but we have made tremendous progress toward building tissue that replicates the form and function of native tissues over just a few decades. The broad clinical application of engineered tissues for treating injury and disease will, in the future, likely have a significant positive impact on human health.

# ORGANS-ON-A-CHIP

Building replacement tissues for the purposes of implantation, as discussed in the previous chapter, generally requires building large constructs in the lab and developing strategies for long-term implantation at a site of disease or damage inside the body. Many researchers, however, focus their efforts on building tissue mimics that are never meant to leave the lab! These tissues, generally only a few hundred micrometers in size, serve an entirely different purpose: they are testbeds for new investigational therapies.

Every medication you have taken, whether an over-the-counter treatment for allergy symptoms or a physician-prescribed antibiotic to treat an infection, involved a tremendous multidisciplinary effort to bring it from the lab to your home. Biologists discover the biochemical mechanisms that are disrupted in a disease, chemists find and manufacture compounds that restore those diseased mechanisms back to their healthy state, and engineers develop methods to deliver those compounds to the right location inside your body at the right time. Drug development typically involves first testing new compounds on living cells cultured in the lab and then testing them on animals that serve as models for humans in healthy and diseased states. The final step in this research process is testing the compounds on humans through federally regulated multistage clinical trials that judge both safety and efficacy. Beyond this, drug development also involves developing methods for sterile and reproducible manufacturing of large quantities of therapeutic compounds in factories. The compounds must then be stored in stable conditions and transported to pharmacies around the world. This process is incredibly expensive, consumes a huge amount of resources, and perhaps most importantly—it takes a very long time.[1]

One of the most time-consuming portions of the drug development process is testing multiple potentially therapeutic compounds for safety and efficacy in the lab. Generating an animal model of a human disease and then studying the effect of an investigational therapy on the progression of that disease involves many experiments over many years. Once a drug is validated in animal models, it is manufactured for preliminary testing in humans. Even after rigorous animal testing, many investigational

Biologists discover the biochemical mechanisms disrupted in a disease, chemists find and manufacture compounds that restore those mechanisms to their healthy state, and engineers deliver those compounds to the right location inside your body at the right time.

compounds fail at the stage of human testing. One reason for this is that the animal models used for initial tests are not perfect models of humans either in healthy or diseased states.[2] Testing several new drugs directly on humans is, however, impractically risky and unethical. Indeed, many people feel that testing investigational drugs on animals is also unethical, but because there is no reasonable alternative for developing the medicines that are crucial to maintaining human life, this remains the standard practice. There are thus many compelling reasons to develop alternative strategies for drug testing that reduce risk and waste while shortening development timelines. One potential alternate strategy is to engineer microscale mimics of human tissues that can be used to test the efficacy of new drugs in a more efficient and ethical manner.

The tissue engineering technologies presented in the previous chapter may seem like the obvious approach to use for this purpose, but building big tissue constructs for the purposes of drug testing would require using many cells, high volumes of cell culture reagents, and large amounts of investigational therapeutic drugs. This becomes very expensive very quickly and is also rather wasteful, as many of the characteristics of a large volume of tissue can actually be captured within a much smaller volume. Using fewer cells per model tissue also means that more models can be generated from the same number of total cells. Having more model tissues allows testing several drugs in parallel, a process termed *high-throughput testing*. It is important to note, however, that a model tissue cannot be made infinitely small. A single cell is often not an accurate representation of a multicellular tissue, in which cell-to-cell communication, as well as interaction between the cells and their extracellular matrix in a 3D environment, are critical to their healthy function.[3] There is thus a middle ground between studying single cells plated on 2D surfaces and studying large-volume tissue-engineered 3D constructs that meets the needs of high-throughput drug testing. This gap in the middle ground is filled by organs-on-a-chip.

In many ways, the considerations of manufacturing 3D cell constructs and maintaining them in culture are the same at the microscale as at the macroscale. Let us revisit liver tissue but look at it from the perspective of developing a liver-on-a-chip that can be maintained and studied in the lab, rather than implanted to combat disease or damage in the body. The liver is particularly interesting when considered from the perspective of drug development, as it is not only relevant for drugs that target liver-specific diseases but also plays a role in the metabolism and toxicity of nearly all drugs introduced into the body. Medicines that induce toxic responses in the liver are, in fact, the leading cause of withdrawals of therapeutic drugs from the market.[4]

Sangeeta Bhatia's lab has developed a liver-on-a-chip model, made using human liver cells, and demonstrated its efficacy in representing the human liver in both healthy and diseased states. In 2008, these researchers patterned small circles of collagen with a diameter of about half a millimeter onto a polystyrene plastic dish and placed primary human hepatocytes onto the circles to form microscale clusters of cells.[5] Surrounding these circles, or colonies of hepatocytes, were fibroblasts derived from a mouse cell line. This dual cell coculture maintained its architecture for over a month, was shown to synthesize proteins and metabolize nitrogen in a manner similar to native liver tissue, and expressed liver-specific genes at a high level. Having established that their microscale liver colonies functioned in a manner resembling the human liver in its physiological environment, the researchers sought to investigate whether this liver-on-a-chip could be used for drug toxicity screening. They observed the effect of over-the-counter compounds like aspirin, as well as prescription medications that had been deemed either safe or toxic to the liver by the FDA. The microscale surrogates for liver tissue demonstrated toxicity profiles similar to those known to occur with native liver tissue, demonstrating that this organ-on-a-chip device could be used as a test platform for drugs prior to human clinical trials. By sorting out drugs with potential negative side effects and thereby mitigating the likelihood that humans are exposed to unsafe investigational therapies, liver-on-a-chip devices have the potential to dramatically improve the safety of clinical trials and reduce the costs of developing new drugs.

One approach to improving the throughput of organ-on-a-chip devices is to construct them within microfluidic platforms, rather than standard cell culture platforms, such as the polystyrene dish described above. Microfluidic devices, described in chapter 2, enable manipulating microliter fluid volumes with extremely precise control. This means that very small volumes of experimental materials, such as investigational drugs loaded into cell culture media, can be used to feed a single tissue, enabling high-throughput screening of how tissues respond to different stimuli in their environment.

These stimuli are not limited to biochemical ones such as investigational drugs. Indeed, the effect of other types of environmental cues, such as physical forces exerted on the tissues, can also be readily tested using such platforms.[6] For example, in 2016, Chris Chen's lab developed a miniature 3D model of connective tissue as a platform for studying how wound healing occurs in the body.[7] The team built a series of microscale wells within a slab of polydimethylsiloxane (PDMS), the flexible silicone most commonly used to fabricate microfluidic devices. Within each of the wells were two microscale pillars. When a mixture of mouse fibroblast cells mixed with collagen were injected into the well, the cells proliferated and exerted mechanical forces on the surrounding gel. These forces caused the gel to compact into a dense tissue-like strip that was tethered at each end by one of the two pillars. The strips served as microscale 3D models of connective tissue found in the body. When connective tissues are injured in their native environment, they tear and can autonomously repair these tears by closing the open gap over time. Chen lab researchers used a microsurgical knife to induce a small tear in their tissue-on-a-chip device and observed how the fibroblasts worked to close the open gap over time (figure 6). They discovered that gap closure was a result of a variety of different mechanisms working in parallel. Fibroblasts started aligning along the boundary of the induced wound, and the tissue started contracting in order to close the gap. The cells also started depositing fibronectin, a protein found in the native extracellular matrix, into the gap to reduce its size, with complete wound closure occurring within a day after damage. This experiment served as a compelling demonstration that organ-on-a-chip devices could be used to study the effect of mechanical cues as well as biochemical cues, lending insight into how we can augment the natural mechanisms by which our bodies respond to damage.

Microscale tissue strips stretched around pillars have been employed to study different types of tissues, such as skeletal muscle, in addition to connective tissue. Skeletal

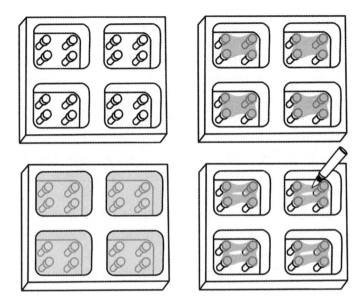

**Figure 6** To form microscale models of connective tissue, researchers formed a series of microscale wells within a slab of flexible silicone (top left). They injected a solution of cells embedded in a collagen gel into the mold (bottom left), which compacted into a dense tissue strip stretched between the microscale pillars (top right). The researchers used a microsurgical knife to tear the engineered tissue (bottom right) and observed how cells remodeled the tear over time. Art by Radha Raman.

muscle helps us generate force and produce motion, and, as a result, disease or damage to this tissue can significantly reduce quality of life.[8] Skeletal muscle contracts in response to an electrical signal from our brains, conveyed via nerve cells known as motor neurons. As muscle is physically connected to our bones via tendons, its contraction drives the motion of bones across articulating joints like elbows and knees. This relatively simple mechanism underlies all the complex motions we can generate, ranging from walking to running to dancing! Disruption of muscle's normal form and function thus results in losses of mobility that can have a severe negative impact on a person's health and autonomy.

In 2009, Nenad Bursac's lab molded a mixture of primary rat skeletal muscle cells (*myoblasts*) embedded in a Matrigel and fibrin gel mixture around PDMS micropillars.[9] The myoblasts proliferated and exerted forces on their surrounding hydrogel matrix to compact into a dense tissue. Once the cells had spread throughout the hydrogel, adjacent cells started to fuse together to form fibers known as *myotubes*, the form in which they become contractile in our bodies. This fusion process was enhanced by tuning the composition of cell culture medium used to feed the cells. Many studies have shown that the fusion of myoblasts into myotubes is promoted by supplementing their culture media with serum from horses. Adding horse serum to their media, therefore, helped the team engineer contractile skeletal muscle tissue. To visualize the internal architecture of the tissues, the researchers used a technique known as immunostaining to tag different types of proteins with different colors of fluorescent dye. They could then use a fluorescence microscope to visually confirm that the individual myoblasts had indeed fused to form aligned myotubes, explaining why the tissue started twitching in culture.

Around the same time, Brian Tseng's lab developed a similar micropillar-based platform for engineering skeletal muscle-on-a-chip devices.[10] Instead of using healthy myoblasts to form the engineered tissues, however, the team used cells from mice that had the same genetic mutation as humans with Duchenne muscular dystrophy, an inherited disease with no known cure that progressively weakens muscle and reduces life expectancy.[11] The cells proliferated and compacted to form microscale tissues, as described above, and stimulating these miniature tissues with an electrical pulse made the muscles contract. Since the muscles were tethered to flexible pillars, their contraction resulted in pillar deflection (figure 7). By recording the tissue with a video camera, the researchers were able to track this stimulation-induced pillar deflection over time. Knowing the size and material properties of the PDMS pillars (e.g., stiffness, stretchiness), they were able to calculate the force that would have been required to drive a deflection of the measured magnitude. This simple

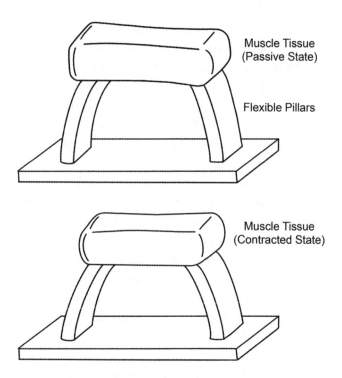

Muscle Tissue
(Passive State)

Flexible Pillars

Muscle Tissue
(Contracted State)

**Figure 7** Engineered skeletal muscle tissue tethered to flexible silicone pillars (top) contracts when electrically stimulated (bottom), resulting in deformation of the pillars that can be tracked over time. The amount of observed deflection is proportional to the force generated by the muscle. Art by Radha Raman.

pillar-based design thus enabled quickly calculating the maximum force a muscle could generate in response to electrical stimulation and served as a simple measure of tissue health and contractility. The team used their muscle-on-a-chip device to test a suite of thirty-one chemical compounds for their ability to treat Duchenne muscular dystrophy. This generated valuable insight into the types, concentrations, and combinations of compounds that helped improve contraction in Duchenne muscular dystrophy skeletal muscle tissue, serving as a step forward in identifying new therapies for combating this debilitating disease.

Many diseases are not caused purely by the malfunction of one cell or tissue type alone but are rather the result of negatively altered behavior across multiple cell types in parallel. A variety of muscular disorders, for example, have origins in the dysregulated function of the motor neurons that control their contraction. It is thus critically important to generate organ-on-a-chip devices that incorporate two or more types of cells, enabling investigating diseases with multicellular origins. In 2016, Roger Kamm's lab designed a microfluidic device in which engineered skeletal muscle tissue and engineered motor neuron tissue were cultured in parallel chambers within the device.[12] The muscle tissue was derived from a mouse cell line, and the neurons were differentiated from a mouse stem cell source. Separating the two tissues mimicked the physical separation between the spinal cord and limbs in our bodies, and the researchers were able to directly observe the neurons growing toward and innervating, or forming a bioelectrical connection with, the engineered muscle tissue (figure 8). Following innervation, the researchers were able to elicit a contraction from the muscle simply by stimulating the neurons, rather than stimulating the muscle directly. The team built on this work in 2018 by replacing the mouse motor neuron cells with iPSCs derived from a patient with amyotrophic lateral sclerosis (ALS).[13] They showed that replacing healthy neurons with diseased neurons resulted in muscle degeneration. However, muscle contraction

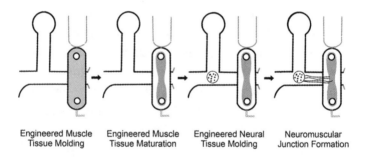

| Engineered Muscle Tissue Molding | Engineered Muscle Tissue Maturation | Engineered Neural Tissue Molding | Neuromuscular Junction Formation |

**Figure 8**  Researchers have used microfluidic devices to generate microscale models of the interaction between neurons and skeletal muscle tissue. They do so by molding muscle tissue within the device (left), allowing it to compact and mature into a dense tissue (middle left), molding neural tissue in a parallel microfluidic channel (middle right), and observing the neurons growing toward the muscle to form bioelectrical connections (right). Art by Radha Raman.

could be recovered through exposure to investigational therapeutic drugs that seem promising for treating ALS. This type of multicellular organ-on-a-chip device could be used, in the future, to understand and test treatments for ALS as well as other types of neuromuscular and neurological diseases.[14]

Multicellular organ-on-a-chip devices also have a significant part to play in the identification and screening of new therapeutics for many other types of diseases. Donald Ingber's lab, for example, made a major advance in this space in 2010 by using two types of cells to model the function of lungs within a microfluidic device.[15] In this device, two microscale fluidic channels were separated by a porous and flexible PDMS membrane. One side of the membrane served as a substrate for culturing human alveolar *epithelial cells*, the cells that form most of the internal surface of our lungs. Human pulmonary microvascular endothelial cells, which line the lung vasculature, were cultured on the opposite side of the membrane. Culturing the two types of cells in two separate chambers enabled separately controlling the biochemical and physical stimuli the cells experienced, such as culture media composition and exposure to air. Since our lungs expand and contract when we breathe air in and out, respectively, the team applied cyclic air pressure to the chambers, resulting in stretching and deformation of the PDMS membrane that mimicked the mechanical action of breathing. This design was used to test whether a multicellular lung-on-a-chip device could serve as an acceptable substitute for native lung tissue. Silica nanoparticles have been shown to have a toxic effect on lungs in animal studies, and the researchers were able to replicate this result in their lung-on-a-chip devices.[16] This experiment served as a compelling demonstration that lung-on-a-chip devices, as well as organ-on-a-chip devices more broadly, could reduce the need for and prevalence of animal testing in the coming years.

In 2012, Ingber's team adapted their lung-on-a-chip device to create a miniature model of the gut.[17] The porous membrane was seeded with human intestinal epithelial cells that, in our bodies, line the small and large intestines, and the channels were continuously perfused with culture medium. Cyclic air pressure changes deflected the membrane to mimic the forces experienced by the gut during peristalsis, the wavelike movements that push food through our gastrointestinal tracts. The engineered intestinal microtissue in the device formed small folds called villi that replicated the structure of intestinal tissue in our bodies. Moreover, culturing a microbe found in the intestine known as *Lactobacillus rhamnosus* GG alongside the epithelial cells was shown to improve the tissue's function as a barrier to small molecules. This effect, which has previously been observed in humans, validated that the gut-on-a-chip device could serve as a useful surrogate for the human gut in both physiological and pathological states.

Ingber's lab has also been responsible for generating a microfluidic model of the blood-brain barrier, the tissue that separates the brain from circulating blood. The blood-brain barrier protects our brains from potentially injurious molecules being transported in the blood. This barrier also makes it hard to deliver therapeutic drugs to the brain, as drugs cannot easily permeate from the blood to the target site of disease or damage. Moreover, dysfunction in the blood-brain barrier has been observed in a wide spectrum of diseases, ranging from brain tumors to Alzheimer's.[18] Understanding the unique structure of the blood-brain barrier within our bodies is thus of significant interest to many biomedical researchers. The Ingber research team manufactured a hollow cylindrical collagen gel within a microfluidic chamber and lined the insides with human brain microvascular endothelial cells, which line blood vessels in our bodies. Culturing other types of brain cells around these endothelial cells, such as human pericytes and astrocytes that are important components of the blood-brain barrier and healthy brain tissue, respectively, resulted in an on-chip mimic of this complex physiological structure.[19] As with the other organ-on-a-chip devices described above, recapitulating a biochemical mechanism known to occur in the human body within this microfluidic platform showed that the device had significant potential as an accurate model for the blood-brain barrier.

Many research teams around the world are working toward recapitulating the function of organ systems inside the body, beyond the tissues mentioned above, inside microfluidic chips. These include work in such diverse and interesting tissues as fat, cornea, bone, skin, and blood vessels.[20] Indeed, as the field progresses, there has been significant interest in connecting different organ-on-a-chip devices together to understand the complex cascade of signaling events that occur between organs in their natural bodily environment.[21] These types of organ-on-a-chip assemblies will likely prove to be even better mimics of the internal dynamics of our bodies, but we have yet to produce model systems that perfectly reproduce the behavior of the entire human body. Such human-on-a-chip devices have not been developed for many reasons, including the fact that we do not yet have a functional model of every human tissue or organ type. Perhaps more importantly, scientists have not been able to develop a single cell culture medium that can simultaneously support the growth and maintenance of several different kinds of tissues in the way that blood does in our bodies.[22] Developing such a medium will likely be one of the critical technical hurdles we must overcome to develop human-on-a-chip devices.

It is important to recognize that not all organ-level functions can be reproduced at the microscale, like the complex macroscale dynamics between our orthopedic tissues (muscle, bone, ligaments, tendons, and cartilage). It is likely that larger tissue-engineered models will

Scientists have not been able to develop a single cell culture medium that can support the growth and maintenance of several different kinds of tissues in the way that blood does in our bodies.

be required to investigate such systems at a macroscale mechanical and biochemical level. Despite this limitation, organ-on-a-chip devices stand to transform the way in which we do basic research. By providing us with a high-throughput way to study the onset, mechanism, and progression of disease and damage in healthy tissues, they provide us with a strong path toward identifying and developing potential cures. Building an organ-on-a-chip device from stem cells derived from a patient could, for example, create a personalized microfluidic model of the patient's disease. This model could be used to prescreen potential medications and dosing regimens for safety and efficacy, increasing the chances that the patient will respond well to a chosen therapeutic plan.

There are a large contingent of researchers that generate microscale 3D tissues using an entirely different approach than that outlined in this chapter. Rather than patterning different cell types into specific 3D architectures, they culture pluripotent stem cells in hydrogels and allow them to mature into different cell types that self-organize into complex 3D architectures. These stem cell–derived self-organizing tissues are referred to as *organoids*, and can be used to study complex tissues in healthy and diseased states. Take neurological diseases, for example, which are the result of biochemical and electrical dys-regulation in distinct regions of the brain.[23] Brain organoids could be used to discover characteristic signatures of neurological disease and serve as testbeds for investigational therapies. A compelling example of a brain organoid, developed by the lab of Li-Huei Tsai in 2016, showed that iPSCs derived from patients with Alzheimer's disease could self-organize into a 3D neural tissue mimic.[24] The brain organoid recapitulated many of the pathological behaviors observed in Alzheimer's patients, such as the aggregation and clumping of specific proteins. The team then showed that some markers of Alzheimer's disease in these organoids could be ameliorated by exposing them to chemical compounds known to reduce the aggregation of clumping proteins. This study, as well as several other pioneering examples of organoids that mimic various systems in the body, has proven that this line of research will have a critical role to play in accelerating new drug discovery efforts. Manufacturing organoids inside the carefully regulated environment of microfluidic devices could, moreover, merge this discipline with organ-on-a-chip research to generate organoid-on-a-chip models.[25]

Regardless of the source of cells used or how they are patterned into 3D tissues, organs-on-a-chip and organoids-on-a-chip could dramatically improve our ability to accurately diagnose and effectively treat human disease. By improving the safety and reducing the cost of developing new therapies, this application of biofabrication has the potential to significantly broaden global access to lifesaving drugs.

# LAB-GROWN MEAT AND LEATHER

Biofabrication has far-reaching implications beyond medicine. A particularly compelling case for building with living cells is in the production of animal-derived food and consumer products, such as meat, dairy, eggs, and leather. Replacements for traditionally sourced meat and leather are the focus of this chapter, as they are the predominant focus of research, development, and commercialization in this field. This is because meat, though a large and important part of many culinary cultures, comes with some potential disadvantages that have been highlighted in recent years. These include the pollution generated by animal farms, the prevalence of foodborne illnesses in animal products, and the ethical concerns raised by housing large and growing numbers of farm animals within diminishing global land resources.[1] Engineered replacements for meat and other animal-derived products have, therefore, become of significant interest to the biofabrication community in recent years.

Most efforts to generate edible biofabricated meat, also termed *cellular agriculture*, have focused on adapting skeletal muscle tissue engineering techniques to this application. While generating large volumes of skeletal muscle tissue is the focus of both tissue engineering and engineered meat applications, engineered meat does not require replicating the microscale architecture and contractile function of native muscle tissue. The four broad goals of engineering meat are to hygienically produce tissue that is fit for human consumption; generate constructs with nutritional values that match or exceed those of naturally sourced tissue; match the macroscale color, texture, smell, and flavor of meat; and rapidly and cost-effectively manufacture large quantities of biofabricated tissue that can be processed, stored, and cooked in a manner that resembles meat farmed from livestock.

Another important distinction that separates lab-grown meat from tissue engineering or organ-on-a-chip applications is that there is a stronger requirement to culture biofabricated meat in serum-free medium, as sourcing serum from animals would reinforce reliance on large-scale livestock farming.[2] Serum replacements derived from other natural sources, such as maitake mushrooms, as well as serum-free culture medias have been developed to address this need and have met with some success.[3]

Skeletal muscle is, of course, not the only component of animal tissue that is used for food production. The intramuscular connective tissue, fat, vasculature, and nervous tissue also play a role in creating the overall form, texture, and taste of meat. A major component of intramuscular connective tissue is collagen, and the distribution of collagen fibers in meat varies across species, from 1% to 15% in cattle to less than 2% in poultry to between 1% and 10% across different types of fish. Collagen fibers in the extracellular matrix are generally wrapped in a matrix of proteins called proteoglycans that bind water and ions such as calcium, sodium, and potassium. Proteoglycans generally make up a smaller portion of muscle than collagen and, in fact, form less than 0.5% of the total dry weight of muscle from cattle. Intramuscular fat is composed of a variety of fatty acids and their derivatives and, like connective tissue, varies in quantity across species. Livestock diet and rearing conditions also have a significant effect on the fat content of tissue.[4] The relative proportions of protein, carbohydrates, and fat determine the nutritional value of the end product and must be carefully regulated in any engineered meat product.

Replicating the nutritional value of native tissue in its live form is, of course, not the primary objective of biofabricated meat. It is equally or perhaps more important to replicate the form and texture of the tissue in its postmortem state.[5] Typically, meat is stored for days or weeks postslaughter and preconsumption, generally in cold or vacuum environments that cause skeletal muscle fibers to shrink to a degree that is dependent on the stress experienced by the animals during slaughter. The extracellular space between cells, by contrast, increases postmortem. Degradation of extracellular connective tissue is considered desirable, as it increases the cooked tenderness of meat. Meat stored under vacuum conditions is purple in color and, when oxygenated, takes on the bright red color that consumers perceive as attractive and fit for consumption. Storage techniques that prevent tissue from oxidizing to brown or black colors, which are negatively perceived by consumers, are very important considerations in meat production. Postmortem changes in texture and color are also affected by changes in nutritional properties, and the rate at and degree to which those changes occur is highly dependent on processing technique. This raises an interesting possibility with biofabricated meat, as there is potential to engineer higher nutritional value into biofabricated products than is possible with traditional meat production. Increasing the concentration of essential amino acids or unsaturated fatty acids without perceptibly changing the texture or taste of meat could, for example, incentivize consumers to opt for healthier choices for themselves and their families.

Hygiene is, of course, also of significant concern in the manufacture and storage of biofabricated meat. Many of

There is potential
to engineer higher
nutritional value into
biofabricated products
than is possible with
traditional meat . . .
[this] could incentivize
consumers to opt
for healthier choices
for their families.

the concerns related to the hygiene of farmed meat, such as the presence of chemical residues from pesticides or other environmental pollutants prevalent on farms, are likely to be mitigated with engineered meat produced in more controlled environments. Concerns regarding bacterial load of the final product and the temperatures required to ensure the product can be safely consumed can also be mitigated by developing manufacturing protocols that precisely determine the composition of and storage parameters for engineered tissues. Accomplishing these goals will, however, involve developing an affordable commercial production process that adheres to global manufacturing standards (e.g., Good Manufacturing Practice—GMP, International Organization for Standardization—ISO) and passes approval by regulatory organizations such as the FDA in the United States.[6]

Important early advances in biofabricating meat have been made by Mark Post and the many researchers trained in his lab.[7] Post's team debuted the first cell culture–derived hamburger, which was cooked and tasted on live television. The researchers faced significant challenges in generating the large volume of tissue required to produce the burger and in differentiating the muscle cells into fibers that resemble mature muscle in animals. In fact, they assembled the burger from 20,000 strands of muscle whose production required several scientists to perform tissue culture protocols thousands of times. As a result, the burger was extremely expensive to produce, costing over $300,000.[8] Calculations by Marianne Ellis and colleagues reveal that producing one kilogram of meat would require culturing 290 billion muscle cells, indicating that manufacturing engineered meat at a meaningful scale requires concurrent advances in bioprocessing technologies.[9] These technologies will not only be relevant for this application of biofabrication but also to others that require reliable, sterile, and cost-effective production of large quantities of living cells, such as the tissue engineering applications discussed in chapter 3.

Post's lab developed a manufacturing technique that used microscale beads as carriers for bovine skeletal muscle cells to scale up their biofabrication process in 2018.[10] Generally, cell culture at a large scale is conducted within bioreactors inside which continuous stirring generates stable flow and mixing. Since many cells, including bovine myoblasts, need to be physically anchored to a substrate in order to live and grow, they cannot be simply grown in these stirred bioreactors, as they would be unable to adhere to any surface. Seeding the cells on microbeads, which can serve as carriers for the cells within the bioreactors, provides a potential solution to this problem. Post and colleagues showed that this method could be used to culture bovine myoblasts in bioreactors and that, as the cells started dividing and proliferating, they transferred from populated beads to empty beads. Differentiating the cells into contractile myotubes required moving the cells off the beads and onto a flatter surface and switching from growth-oriented to differentiation-oriented culture media, as noted in chapter 4. This can be done by separating the cells from the microscale beads, dissolving the beads, or making the beads edible so they can be embedded in the end product. Post and colleagues speculate that the latter approach is likely the optimal approach, as it would reduce the number of physical and chemical processing steps required to complete the production process. Moreover, edible beads could be engineered to have advantageous properties that enhance the flavor, texture, and nutritional value of the final product.[11]

Other researchers in this space are testing the potential of growing cells in a scaffolding formed of soft fibrous materials rather than microbeads. This could potentially reduce the complexity of the harvesting process, as there would be no need to transition cells from microbeads to flatter surfaces to differentiate them into mature contractile myotubes. In 2019, Kevin Kit Parker's lab developed a method for rapidly manufacturing gelatin microfibers using food-safe solutions and solvents with a technique called immersion rotary jet spinning.[12] The team chose gelatin as a substrate for engineered meats as it is derived from collagen and could replicate some of the biochemical cues found in natural meat. Moreover, manufacturing it in a fibrous form could also recapitulate some of the natural physical structure of the extracellular matrix environment in native muscle. Seeded within these pig-derived gelatin fibrous scaffolds were two types of cells: aortic smooth muscle cells from cows and skeletal myoblasts from rabbits. Both cell types were shown to attach to the fibers and aggregate into clumps on short fibers around twenty micrometers in length. Longer fibers, by contrast, drove cells to align along the longitudinal axis of the fibers.

In this preliminary work, the Parker lab's researchers were not able to recreate the mature contractile architecture that helps skeletal muscle contract. Their biofabricated meat tissues thus more closely resembled processed meats such as ground beef, rather than steak. As the cell lines used to manufacture this tissue have limited capacity to differentiate, it is possible that choosing a different cell source or developing a more efficient differentiation protocol could help improve the texture of the engineered tissue. As protocols to differentiate iPSCs from animals into relevant types of tissues (skeletal muscle, fat, vasculature) improve, iPSCs could become a promising source of raw material for biofabricated meat in the future.

Encouragingly, the engineered tissue produced by Parker's team generated collagen and collagen-like protein expression at similar levels to natural rabbit skeletal muscle and bacon. The engineered tissue was maintained in culture for three to four weeks and, when tested for mechanical properties using techniques standard to the food industry, demonstrated similar properties to ground beef, which increases in hardness after cooking. By contrast, tissues with differentiated mature muscle fibers, such as fresh rabbit muscle and beef tenderloin, were shown to soften after cooking. As discussed above, cooking naturally derived meat can result in collagen degradation that causes the meat to become more tender. It is possible that altering the composition of the gelatin fibers used by Parker's team to something that partially solubilizes when cooked could help further soften the engineered meat, recreating the properties of naturally derived meat. Fibrous scaffolds could also be fabricated using other food-safe materials, potentially derived from plant sources, that could prove more nutritious than gelatin.

One such alternative scaffold material being investigated by Shulamit Levenberg and colleagues is textured soy protein, an inexpensive edible byproduct of soybean oil processing.[13] In 2020, this team showed that skeletal muscle cells derived from cows could be seeded within soy scaffolds and differentiated to form skeletal muscle fibers. Interestingly, culturing skeletal muscle cells alongside smooth muscle cells (which form our involuntary muscles such as the stomach) and endothelial cells (which form our blood vessels) was shown to improve tissue formation and increase the cells' ability to deposit extracellular matrix proteins. This coculture approach also produced tissues with mechanical properties that better resembled those of natural cow muscle, and taste tests of the team's engineered meat generated positive feedback in terms of both flavor and texture.

Others have also studied coculture platforms for engineering meat but have focused instead on growing skeletal muscle cells alongside adipose, or fat tissue. Intramuscular fat improves meat texture and juiciness, and producing engineered meat composed of both skeletal muscle cells and fat cells will be critical to replicating these qualities of traditionally sourced meat. Researchers in Takanori Nishimura's lab tested the effects of coculturing fat cells (*adipocytes*) and skeletal muscle cells in a 2D culture platform and discovered that the presence of fat tissue suppressed the growth and development of the muscle tissue into contractile fibers.[14] They attributed this to the biochemical signals emitted by the fat cells, which disrupted the maturation of the muscle cells. Disrupting this cross talk between the two types of cells will be critical toward integrating engineered fat and muscle into biofabricated meat. This could be done either by constantly cycling in fresh media that removes the biochemicals secreted by the fat cells or by culturing the muscle and fat tissues separately and only combining them during the final stage of manufacturing.

In addition to experimenting with different types of cells, researchers have also studied engineering meat from various sources beyond mammals or fowl. Insect-derived cells have, in particular, been of interest to the engineered food community, as they are generally easier and cheaper to produce at large scales and are more tolerant of broad ranges in environmental conditions such as pH and temperature. Established protocols for serum-free culture of insect cells make insect-derived meats an especially appealing prospect from a manufacturing and sustainability standpoint. However, consumer acceptance of insect-based foods may prove to be one of the most significant barriers to commercial translation of this technology. Population surveys have revealed that most people are unwilling to replace animal-based meat with insect-based meat and that this is particularly the case in Western countries where this source of nutrition is unfamiliar.[15]

It is important to note that even engineered meat from mammals or fowl may generate a negative reaction from many consumers and slow or stop its adoption. A study by Courtney Dillard and Christopher Bryant in 2019 revealed that the manner in which biofabricated meat is framed by the media can have a significant impact on public perception.[16] The researchers presented study participants, all of whom were adults aged twenty-six to thirty-five living in the United States, with one of three potential motivations for engineering meat: lab-grown meat is more sustainable and ethical, lab-grown meat uses cutting-edge technology, or lab-grown meat tastes just like regular meat but can be healthier. The participants were then asked how they felt about this new technology, and the study found that about 65% were willing to try engineered meat and that men and younger people were more likely to view this technology positively. The researchers also found that vegetarians and vegans were less willing to try engineered meat than those who incorporated meat and fish into their regular diets. As a lifelong vegetarian myself, I find this quite relatable, as my diet already revolves around plant-based foods, and engineered meat would not be substituting for an existing component of my diet. Perhaps the most interesting outcome of this study was that participants who were presented with the motivation of engineered meat being high tech had the most negative attitudes toward the concept. As this is the framework most often used in media coverage of lab-grown food, Dillard and Bryant suggest that changing the tone of public-facing content will be critical to increasing consumer adoption of biofabricated meat.

Presenting engineered meat as a source of culinary innovation could assist with marketing efforts to increase widespread consumer adoption. After all, engineered meat can do more than just recover the form and taste of farmed meat. Indeed, leveraging novel materials and manufacturing techniques opens up the possibility of designing food that cannot be created using traditional culinary approaches. The technology of 3D printing, and more specifically the extrusion-based 3D printing approach described in chapter 2, has generated significant interest

Consumers were asked how they felt about this new technology and about 65% were willing to try engineered meat and men and younger people were more likely to view this technology positively.

in the engineered food industry, as it enables creating multi-ingredient 3D structures with complex internal architectures. Indeed, several printers developed by both research labs and commercial food entities have demonstrated the ability to build with chocolate and cheese and even make a custom pizza!

Hod Lipson's research lab has made some significant advances in the food printing space, but a recent innovation from the lab, published in 2019, is of particular interest in the context of engineered meat.[17] The team combined an extrusion 3D printer with an infrared heating source to cook food layer by layer as it is printed. Infrared generates uniform heat and can be used to dry, bake, roast, or broil food in a targeted and precise manner. The team developed sesame-based, chicken-based, shrimp-based, and dough-based pastes and used rice flour to tune the firmness of the pastes until they resembled the texture of cream cheese, enabling facile printing. Since these materials were composed of raw meat and eggs, they needed to be cooked prior to consumption. Lipson's lab combined the sesame, chicken, and shrimp pastes to manufacture a pate bite printed in the shape of a twisted pyramid. Infrared heating was used to cook the deposited pastes, and changes in color and texture of the printed material after heating validated that cooking was occurring. This approach is particularly interesting as a method for manufacturing multimaterial foods, as being able to control the degree of heat applied to different regions of a printed food allows mixing ingredients, textures, flavors, and colors in a more creative manner than possible with traditional cooking. While the adaptability of 3D printing to large-scale manufacturing of consumable meat is presently unknown, it is clear that biofabrication technologies can enable exciting next-generation innovations in cuisine.

Beyond consumable meat, biofabricated tissue for nonmedical applications also has significant applicability in the production of consumer goods such as leather. Many consumer products are fabricated from leather, ranging from furniture to purses to shoes. Natural leather has its origins in the skin of animals, which goes through a series of intense chemical processing steps to remove hair, flesh, fat, blood vessels, and nonstructural proteins. This process strips cells from the animal hide and leaves behind the collagen extracellular matrix. Gabor Forgacs and his team have developed a method that relies on biofabrication to produce a leatherlike engineered material.[18] To recreate the microscale structure and macroscale texture of leather, the team reverse engineered natural leather by culturing 2D monolayers of bovine skin fibroblasts. These primary cells could be sourced from a biopsy of the skin, causing the cow minimal harm, and naturally produced collagen. Once a 2D cell layer secreted enough collagen, a second layer of cells was stacked on top of the first, and the process was repeated sequentially until the desired thickness was achieved. A trilayer of cells fused together could be harvested after eighteen days in culture.

The researchers made a mimic of animal hide by layering five trilayers on top of one another to form a fifteen-layer construct. They then tanned the hide using a variety of different techniques adapted from traditional tanning but without the preliminary steps required to strip the tissue of hair and flesh, such as aluminum salt-, vegetable-, chromium-, and aldehyde-based techniques. The engineered hide was colored and then fixed with fat liquor, a fat–soap mixture that introduces oil into the tissue, followed by washing and air-drying. Comparing the engineered leather mimic to hide from sheep, kangaroos, and goats showed that the tissue looked fairly similar and somewhat more uniform than naturally produced leather. However, engineered leather tore in response to the stitching techniques used for commercial manufacturing of animal-derived leather. The researchers attributed this difference in material strength to the fact that the engineered leather was thinner than the animal-derived leather, and that the dead cells had not been removed from the tissues. Moreover, skin fibroblasts may not yet be able to produce collagen as efficiently in a lab setting as in a living animal. As a result, collagen content was much lower in the engineered leather, forming 30% of the material as opposed to 80% of natural bovine leather.

Regardless of these differences from natural leather, the engineered leather was still stretchy and workable, and the team was able to fashion it into a bracelet! In addition to having a very uniform finish, the amount of leather generated to build the bracelet or any other consumer product could be predetermined before manufacture, ensuring minimal waste. Engineering animal hide could also enable, in the future, designing new types of leatherlike materials with interesting and advantageous properties not found in nature. While the material developed by Forgacs's team cannot officially be called leather, as it is not derived from the skin of an animal, engineering a tissue that has the look and feel of the material consumers desire and associate with luxury goods can have a significant impact on commercialization. Consumer choices are driven by many factors, ranging from price to quality to ethical and sustainability considerations, and it seems likely that we will all have to make a choice between engineered mimics and naturally derived materials in the coming years. The economics and ethics of these choices are discussed further in chapters 7 and 8.

There are many motivations to explore alternatives to producing consumables from livestock. The rapidly growing global population has led to an increase in demand for nutritious and filling food, and large swaths of land must be dedicated toward breeding and maintaining animals and harvesting their meat for food to meet this demand. There is a limit on the amount of land that can be dedicated to this cause, and animal farming land will become an increasingly scarce resource in the face of competing demands for living space and farming of grains and vegetables. This has generated widespread concern about our inability to meet the growing demand of meat consumption in the coming years, alongside increasing worries about the negative environmental impact of large volumes of livestock and their correspondingly large carbon footprint. Moreover, many have argued that crowding animals can lead to unhealthy and unsafe breeding conditions, triggering ethical dilemmas about this method of producing consumable meat.

It is clear, therefore, that a significant change in the technology of meat production will likely be required to generate safe, affordable, ethical, and environmentally conscious alternatives or supplements to traditional animal farming. Beyond these advantages are the potential health benefits associated with lab-grown meat. Since traditionally sourced meat, especially red meat and processed meat, may increase the risk of diseases such as heart disease and colorectal cancer, it could prove advantageous to develop engineered meats with reduced concentrations of the biochemicals that play a role in disease onset and development.[19] Engineering customized edible products with biofabrication tools could thus enable boosting nutritional value and limiting adverse health effects for a range of animal-derived foods.

It is important to note that while the cellular agriculture approaches described in this chapter are in early stages of development, the field is growing and changing rapidly. As of the end of 2020, engineered meat has been approved for human consumption for the first time in Singapore, and it is possible that it may become more broadly available soon.[20] The regulatory discussions surrounding the safety of this technology, as well as its environmental and economic impact, are happening now, as further discussed in chapter 7. These discussions, coupled with technological advances in developing serum-free culture media and scaling up tissue manufacturing, will enable scientists to truly leverage the potential of biofabricated food and consumer products and maximize potential impact on our economy and environment.

# BIOHYBRID MACHINES

Thus far in our exploration into biofabrication, we have discussed different avenues by which we can recreate the structure and function of biological systems that are already present in nature. Our goal has been to replicate the conditions generated by our natural environment as closely as possible to recreate tissues and organs for medicine, food, and consumer products. This was indeed the focus of biofabrication technology development and applications in the initial years and decades of its scientific rise. However, some biofabrication researchers have now started to move away from reverse engineering what already exists in our natural world and toward forward engineering biofabricated constructs with nonnatural or hypernatural functionalities.

Why would forward engineering with biological materials generate value? The responsive nature of biological materials, as discussed in the introduction to this book, are unmatched by the abiotic materials with which engineers typically build. As a result, incorporating biological components into machines could yield engineered systems with unparalleled ability to dynamically sense and adapt to the needs of changing environments. These new part-biological and part-synthetic biohybrid machines could augment the existing tools and technologies we rely on to improve human health, productivity, and quality of life. The scientific community is still grappling with what a forward-engineered biohybrid machine would look like, or what it could do, but has started exploring how biofabrication could impact a specific subset of machines: robots. Broadly defined, robots are machines that autonomously sense, process, and respond to cues from their surroundings in real time. A robot that uses biological tissues to accomplish one or all of these functions can be called a biohybrid robot. Biohybrid robots form a very new branch of the discipline of biofabrication, but research in this field has grown and diversified rapidly within the last few years.[1]

Most of the research in biohybrid robotics has focused on using biological tissues as actuators, the components in machines that generate force and produce motion. There are many types of synthetic actuators that power the machines we use every day. Electric motors, for example, are used in computers, exhaust fans, refrigerators, and

Part-biological and part-synthetic biohybrid machines could augment the existing tools and technologies we rely on to improve human health, productivity, and quality of life.

cars. These actuators function by converting electrical energy into mechanical energy, but other types of commonly used actuators such as shape memory alloys, pneumatic cylinders, and hydraulic cylinders convert energy from heat, air pressure, or fluid pressure, respectively, into mechanical force output. Our bodies also use actuators, albeit of a biological origin, for a variety of functional activities such as generating our heartbeats, moving our limbs, and digesting the food we eat. The tissues we use to perform these functions, namely cardiac muscle, skeletal muscle, and smooth muscle, have evolved to generate complex coordinated forces very efficiently and in ways we have yet to replicate with synthetic materials. In fact, skeletal muscle can produce larger forces from smaller volumes than nearly any synthetic actuator and can also do things that synthetic actuators cannot, such as getting stronger in response to exercise or healing in response to damage.[2] In recent years, many researchers have been inspired by the exceptional functional capabilities of living muscle, as compared to traditional synthetic actuators, and grown curious about whether it can be harnessed to power motion in engineered machines.

The first studies to investigate this idea relied on directly removing intact living muscle tissue from an animal, coupling it to a robot skeleton made of a synthetic material, and using it to power actuation of the skeleton. A pioneering study in this field, performed by Hugh Herr and Robert Dennis in 2004, tested this approach for designing a biohybrid robot.[3] The team extracted two skeletal muscles with intact tendons from frogs and sutured the tendons to either side of a thin ellipse-like substrate made of PDMS and other polymers. In the body, skeletal muscle receives an electrical impulse from motor neurons that tells it when to contract. To replicate this nerve signal, the researchers designed an on-board electrical stimulator that drove current through wires that were wrapped around each muscle. The resulting biohybrid robot was placed in a watertight tank containing Ringer's solution, a mixture of several types of salts dissolved in water that is often used to preserve tissues or organs extracted from living animals. Electrical stimulation of the muscles using the on-board stimulator made the robot swim around the tank, with a speed that could be tuned by optimizing the rate and degree of electrical stimulation.

While this was a fascinating demonstration and a critical first step toward proving that robots could be powered by biological tissues, using muscle tissue directly removed from animals is not an ideal approach for a few reasons. There is a significant amount of variation between individual animals in a species, so the exact size of the muscle or force generated by the muscle would differ widely between engineered devices depending on which animal supplied the tissue. Moreover, building each device would require sacrificing an animal, which raises ethical concerns as well as sustainability concerns. Being able to biofabricate muscle actuators in the lab to power biohybrid robots would mitigate these concerns and would also provide greater flexibility and control over the size, shape, and functional output of biological actuators.

In 2007, Kevin Kit Parker's lab took a significant step toward this goal by exploring the use of biofabricated cardiac muscle thin films as actuators.[4] While the basis of cardiac muscle contraction is also electrical, those electrical signals are not consciously generated by our nervous system but rather spontaneously generated by the heart's pacemaker, known as the SA node. Another important difference between cardiac and skeletal muscle is that cardiac muscle cells (cardiomyocytes) are electrically connected to one another, to ensure that they beat in synchrony. When you consider that cardiac muscle has evolved to keep our hearts beating, you can see why contraction occurring at the same time throughout large portions of the tissue is critically important for healthy biological function. Knowing these underlying principles of how cardiac muscle is organized and controlled, Parker's group decided to make a swimming robot powered by cardiomyocytes.

The team manufactured thin flexible films (less than sixty micrometers in thickness) of the polymer PDMS and coated these films with fibronectin, an extracellular matrix protein that binds to other extracellular matrix proteins like collagen and fibrin. They cultured primary heart muscle cells derived from rats on these coated films. When the 2D films were released from an underlying substrate, they could adopt different 3D conformations depending on the geometry and material properties of the PDMS. As the cardiomyocytes spontaneously started to contract in synchrony, they generated large cyclic forces at a frequency that could be controlled using an externally applied stimulus. The researchers used this fabrication technique to explore how different 3D shapes formed from cardiomyocytes and PDMS films could be used to power a wide range of motions.

Long rectangular strips of PDMS with cardiomyocytes aligned along the length of the strip and on the concave side of the strip generated cyclic contractions from an uncoiled to a coiled state in response to muscle contraction. By contrast, patterning the cardiomyocytes along discrete lines on the convex side of the film generated a helix shape that contracted along its long axis. Both these designs mimicked functional actions, such as peristaltic or suction pumping, that can be useful for a variety of machines. Some functional applications require discrete rather than cyclic actuation, and the researchers leveraged the biological concept of tetanus to achieve this function. Physiological tetanus (not to be confused with the disease tetanus) is a state of sustained muscle contraction that occurs when muscle is stimulated at a very high frequency, such that it cannot sufficiently relax between sequential contractions. For the cardiac cells used by this group of researchers, tetanus occurred at a frequency of five hertz, or five contractions per second. They manufactured a long, thin ellipse-shaped strip of PDMS and cardiac muscle, which opened when the muscle was relaxed and closed when the muscle was contracted. This strip could be electrically paced at sequentially higher frequencies until tetanus was accomplished, forming a stable grip between the two edges of the film. This cardiac muscle–powered gripper could potentially be useful for carefully handling small soft items, such as cells from biological samples. Beyond these applications in pumping and gripping, the research team also engineered walking and swimming robots powered by the contraction of cardiac muscle. The walking design, composed of a passive front leg and a cyclically contracting rear leg, could crawl along the bottom of a petri dish filled with culture media. The swimming design, which took the form of a triangle, could propel itself in one direction through a liquid culture medium.

Building on this initial work, Parker's team presented a further advance in 2012 by using design principles from living organisms to develop biohybrid robots capable of more complex and lifelike motions.[5] They mapped out the architecture of the muscle cells that scyphozoan jellyfish use to swim and learned that the cyclic propulsive motion generated by such jellyfish requires a very specific cellular layout. Highly aligned muscle fibers must contract with precise synchrony in order to generate the types and magnitude of forces observed in jellyfish in their natural state. The two-component design described above, of cardiomyocytes coupled to a PDMS thin film, was explored as a platform to mimic this behavior. The body shape of the jellyfish, composed of a central lobe with surrounding fins, was manufactured using PDMS. Cardiomyocytes were aligned along the fins and connected to adjacent fins via a circumferential ring in the lobe, ensuring synchronous contraction of the entire jellyfish-like robot body. When these biohybrid robots were placed in a media bath and paced with external electrical stimulation, they demonstrated propulsion and swimming behavior similar to the actual jellyfish on which they were modeled (figure 9)! This served as a proof that biological design principles, such as how bodies are designed and muscles are organized, in addition to biological materials themselves, could be used to build biohybrid robots that move just like some of the organisms in our natural world.

Other researchers have also used the concept of cardiomyocytes coupled to PDMS skeletons to generate complex motions from biohybrid robots of different shapes and sizes. Taher Saif's lab demonstrated, in 2014, that a variation on this approach could be used to power robots that swim like spermatozoon, more commonly referred to as sperm.[6] They fabricated PDMS in the form of a spermatozoon, composed of a small stiff head with a long

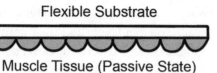

Flexible Substrate

Muscle Tissue (Passive State)

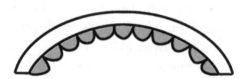

Muscle Tissue (Contracted State)

**Figure 9** A biohybrid robot is formed by coupling a layer of cardiac muscle tissue to a flexible silicone substrate. As the cardiac muscle contracts, it deforms the flexible substrate, resulting in jellyfish-like propulsive swimming behavior. Art by Radha Raman.

thin flexible tail. The team seeded between one and three primary rat cardiomyocytes near the side of the tail that formed a juncture with the head, and the resultant muscle contraction generated bending waves that propagated along the length of the flexible tail, generating a swimming motion from the miniature device. Interestingly, incorporating two tails rather than a single tail in the device led to a dramatic increase in the swimming speed of the robots from around ten micrometers per second to over eighty micrometers per second. This showcased that being able to control various aspects of the design of a biohybrid robot using biofabrication techniques enabled high-level control over device function. Saif's lab showed that their devices could maintain consistent muscle-driven actuation for an average of three to four days in culture and hypothesized that incorporating more complex multitail designs or other cell types into their swimming robot could power even more interesting functional behaviors in the future.

In 2013, Ali Khademhosseini's lab explored a different substrate for cardiomyocyte-powered biohybrid devices by replacing PDMS with gelatin methacrylate hydrogels.[7] Embedded within these hydrogels were electrically conductive fibrous networks formed from carbon nanotubes (CNTs). This fibrous substrate provided physical cues that helped cardiac cells adhere to them and form electrical connections with one another. As a result, cardiac cells seeded on the conductive CNT hydrogels demonstrated more stable spontaneous contraction as compared to cells seeded on plain hydrogels. To demonstrate that CNT-hydrogel substrates could be used to produce functional motions, Khademhosseini's group formed their cardio-myocyte and CNT-hydrogel platform into a concave rect-angular strip. They demonstrated that cyclic contraction of the heart cells drove opening and closing of the strip from a loosely rolled up tube into a more tightly coiled tube. This type of cyclic pumping action could prove useful in many real-world functional applications, as described above. The team also produced a triangle-shaped swim-mer device that could propel itself in culture medium in response to an external electrical stimulus.

Interestingly, the presence of CNTs in the substrate was shown to help reduce or prevent the damage induced by chemical compounds that are normally toxic to cardiac cells, such as the chemotherapy drug doxorubicin. When cardiac muscle cells seeded on plain hydrogels were sub-jected to this toxin, which was introduced into the sur-rounding culture media, they experienced a disturbance in their ability to contract. Cardiac muscle cells seeded on CNT-hydrogels, by contrast, were able to resist this contraction disturbance. One potential explanation for this behavior was that the fibrous networks of electri-cally conductive CNTs were helping propagate electrical signals between cardiomyocytes in the device, even when they were not able to communicate with each other using their normal biological mechanisms. This is one of many experiments by this team and others that showcases the advantages of leveraging the best of both biological materials and synthetic materials in the design of biofabricated machines. Indeed, it seems very likely that both synthetic and biological components will play active roles in the next generation of biohybrid robots.

While cardiac muscle generates significant and synchronous forces, it is not the form of muscle that most animals use to accomplish complex functional behaviors such as locomotion. Skeletal muscle is the primary driver of voluntary force production in our bodies and the bodies of other animals. It possesses many advantageous qualities that render it particularly well suited to this task. Skeletal muscle tissue is able to generate large forces from small volumes. It has a hierarchical structure composed of many contractile myotube fibers acting in parallel to generate aligned and coordinated forces. This enables muscle to dynamically adjust its form and function to both positive stimuli (such as exercise) and negative stimuli (such as damage) by forming more new muscle fibers on demand. These capabilities are, as yet, unmatched by the synthetic actuators traditionally used by engineers.

Unlike cardiac muscle, which contracts involuntarily and is always beating, skeletal muscle's contraction can be turned on or off on demand. In the body, electrical signals conveyed to muscle through motor neurons drive on-demand contraction. Replicating this external control in a lab setting enables treating skeletal muscle tissue just like any other functional component in an engineered machine: something that can be turned off and on and told how much force to produce, with precise spatiotemporal control. Unlike a traditional functional component made out of an abiotic synthetic material, however, this biological material dynamically adapts to its surroundings. This has served as a compelling motivation to investigate skeletal muscle as an actuator for biohybrid robots and formed the basis for my doctoral research in the lab of Rashid Bashir.

In 2014, our team biofabricated a skeletal muscle–powered robot in the lab.[8] To do so, we took inspiration from our bodies, in which skeletal muscle is attached to bones via tendons and stretched across articulating joints, such as elbows or knees. When the muscle receives a nerve stimulus to contract, it does so and is able to generate a wide range of joint motions. We imitated this native architecture in the lab by 3D printing a polymer skeleton in the shape of a flexible beam (mimicking an articulating joint) connecting two stiff pillars (mimicking tendons). We placed this skeleton in a miniature well and injected a solution of skeletal muscle myoblasts and natural hydrogels that resembled the environment surrounding these cells in the body. This cell-gel mixture solidified, as described in chapter 2, and the cells within it replicated themselves and exerted mechanical forces on the surrounding gel. Over time, the cell-gel mixture compacted into a dense 3D engineered tissue strip. Earlier protocols developed by other researchers had shown that a few alterations to the cell culture media, such as switching from bovine-derived serum to horse-derived serum, would stop the cells from replicating themselves and instead encourage them to fuse together. We replicated this protocol with our devices and observed the embedded myoblasts fusing together to form myotubes, the basic contractile unit of skeletal muscle.

When we placed the resulting biohybrid robot between two electrodes and sent an electrical pulse through the tissue, we were able to evoke on-demand muscle contraction that deformed the robot's polymer skeleton. Furthermore, as the skeleton was designed to be asymmetric, with one pillar being a little longer than the other, each muscle contraction resulted in a larger deflection of the longer pillar as compared to the smaller one. As a result, the robot crawled in the direction of the longer leg, in a manner similar to an inchworm caterpillar crawling along a twig.

Interestingly, the speed of the walking robot could be controlled by a few different mechanisms. Increasing the frequency of electrical stimulation, from one pulse per second to four pulses per second, for example, generated a corresponding increase in the number of muscle contractions per second and a resultant increase in robot speed.

Increasing the frequency too high, however, to ten pulses per second resulted in muscle tetanus, as described earlier. We discovered that robot speed could also be tuned through other mechanisms. For example, adding biochemicals that promote skeletal muscle growth, like human insulin-like growth factor (IGF-1), to the cell culture media resulted in a significant increase in the force generated by the muscle. We also observed that making the polymer skeletons stiffer imposed a mechanical stretch stimulus on the muscle tissue, helping it achieve better organization and maturity and generate larger contraction forces.

These experiments showed us early proof that machines that were part biological could not be thought of in the same way as machines that were purely synthetic. No synthetic actuator dynamically adapts the force it produces in response to a changing mechanical or chemical environment, but the same was clearly not true for our skeletal muscle actuators. Moreover, by using myoblasts from a cell line, rather than a primary cell source as with prior biohybrid robots, we showed that a more sustainable and reliable approach could be used to manufacture muscle-powered robots.

To build on this work, our team set about making two design changes to our skeletal muscle–powered robots.[9] First, instead of molding the engineered muscle tissue as a strip that formed around the polymer skeleton, we decided to mold it in the shape of a rubber band. These muscle rubber bands (about six millimeters in length) could be picked up with a pair of sterile tweezers and stretched around any of a wide variety of 3D printed polymer skeletons. This simple design modification meant that, without changing the muscle tissue form factor, we could easily adapt it as an actuator for more than just walking robots. Indeed, researchers from Taher Saif's lab has since shown that placing a skeletal muscle rubber band around a hollow flexible tube can be used to make a muscle-powered peristaltic pump.[10]

The second design change we implemented was to use genetic engineering tools to make the skeletal muscle cells contract in response to a pulse of light, rather than just a pulse of electricity. Many other researchers have worked toward making cells, specifically neurons, responsive to light in recent years, and the field as a whole is called optogenetics. By implementing this approach in muscle cells, we showed that flashing light on engineered muscle tissue could make it contract in much the same way as pulses of electricity. This is advantageous because a pulse of light can be spatially focused in a way that is very difficult to do with electricity. In other words, you can shine light on just the portion of tissue that you want to contract, rather than making the entire muscle contract at the same time. In a completely symmetric two-legged biohybrid robot, for example, sending an electrical pulse through the culture media would activate both legs at the same time.

This would result in symmetric contraction with zero net movement. Shining light on only one leg, however, results in contraction of only that muscle, which then drives the robot to walk in the direction of the stimulated leg (figure 10). Similarly, shining light on only half of a muscle can make the robot turn and rotate. Genetically engineering muscle to be light-responsive, therefore, resulted in

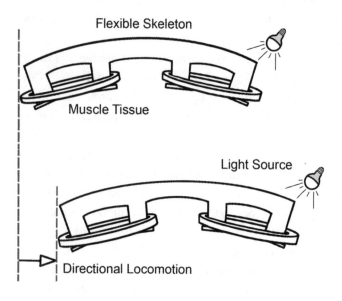

**Figure 10** A biohybrid robot is formed by coupling two skeletal muscle rubber bands to a flexible skeleton. The muscle is engineered to contract in response to a pulse of light. Shining light pulses on the rightmost leg of a symmetric two-legged robot thus results in walking, or directional locomotion, of the robot toward the right. Art by Radha Raman.

biohybrid robots that could move in any direction on a 2D surface on demand.

Other researchers in this field have also shown that optogenetic techniques can be used to generate light-guided biohybrid robots. In 2016, Parker's team developed a cardiac muscle–powered light-responsive robot, inspired this time by a stingray rather than a jellyfish.[11] Replicating the stingray's undulating fin movement required a slightly more complex design than simply reproducing the stingray's body shape using a PDMS thin film. Rather, the undulatory motion of this fish requires a muscle layer that produces downward contraction as well as a second layer of muscle that produces upward contraction. Since the second layer is essentially the antagonist toward the first layer's contraction, the team hypothesized that its action could be replicated using a stiff gold backbone that is deformed during a downward contraction and whose elastic energy helps it rebound to its original shape. The final skeleton design was thus composed of a gold backbone layered between two thin films of PDMS, on top of which was seeded a sheet of cardiomyocytes.

Undulatory contraction occurs when muscles contract in a wave that propagates across adjacent cells, so recapturing this movement required a way of spatially controlling when and where the cardiac cells contracted along the body of the robotic stingray. The researchers addressed this by genetically engineering the cardiac cells to respond to blue light, as we had done with skeletal muscle. They used spatially focused light pulses to generate sequential waves of muscle contraction that propagated along the length of the stingray from front to back. This helped the biohybrid robot replicate the stingray's swimming behavior. Moreover, as this technique generated a robot that followed the light, they were even able to make their biohybrid stingray maneuver its way around an obstacle course in a liquid tank!

These mammalian cardiac muscle– and skeletal muscle–powered devices are just a few examples of many fascinating demonstrations of biohybrid robots that have emerged in recent years. Pioneering examples of nonmammalian tissue–powered devices, such as robots powered by tissue from sea slugs, inchworms, and bacteria, have significantly broadened and deepened our knowledge of how to build robots using biological materials.[12] It is important to remember, however, that all of the research outlined in this chapter is at a very early stage. Showing that robots can use biological actuators to move around certainly is not the same as showing that biological actuators are better than abiotic synthetic actuators in some measurable way. There are a few promising indications, however, that this could be the case.

Many tissue engineering researchers had demonstrated in the past that exercising engineered skeletal muscle tissue, by repeatedly stimulating it with electrical pulses for example, resulted in increasing the tissue's contractile force.[13] Our team in the Bashir lab showed that exercising optogenetic muscle-powered robots with repeated light pulses over several days also resulted in the engineered muscle getting stronger over time.[14] Moreover, we showed that robots that experienced some form of damage, like tearing of the muscle actuators, could be healed by the addition of new healthy cells to the damaged tissue.[15] Traditional robots, powered purely by synthetic materials, have yet to be able to do things like exercise to get stronger or heal from damage.

While these developments are promising, significant technical challenges will need to be addressed before biological robots can be more broadly applied to real-world technical challenges. For example, researchers must be able to integrate other types of functionalities beyond actuation into biological robots. Using different types of cells, such as using blood vessels for transporting nutrients or neural networks for processing stimuli and controlling output, could be one potential path toward addressing this critical need, but it is also possible that these functionalities could be generated using synthetic materials that interface with biological actuators.[16] Using biological robots in different environments might also require developing new methods to freeze, store, and transport biohybrid machines, or building exoskeletons that can protect the robots from harsh or unpredictable surroundings.[17]

Traditional robots, powered purely by synthetic materials, have yet to be able to do things like exercise to get stronger or heal from damage.

Despite the remaining technical challenges in biohybrid design, the early studies in this field prove forward engineering with biological materials is possible and can provide a functional advantage over building with abiotic synthetic materials alone. The biohybrid robots presented in this chapter are the first demonstrations of their kind and show significant promise for future applications of biofabrication in machines.

# ECONOMIC AND ENVIRONMENTAL IMPACTS OF BIOFABRICATION

The scientific advances that enable biofabrication, as well as the technical challenges that remain to be addressed before mass commercial adaptation, are not the only areas of impact that must be considered in a holistic analysis of this field. Despite how potentially advantageous biofabricated advances in medicine, agriculture, robotics, or consumer goods may be, their mass adoption can only occur when we can produce them at scale and in a cost-effective manner. Moreover, ensuring biomanufacturing processes can be conducted in a manner that reduces environmental impact, as compared to current methods, will be an important consideration as we move toward a more sustainable future. Developing affordable and environmentally friendly biofabrication approaches is the only way we can ensure this next-generation technology can be leveraged by the vast majority of the global population for many years to come.

Developing affordable and environmentally friendly biofabrication approaches is the only way we can ensure this next-generation technology can be leveraged by the vast majority of the global population for many years to come.

Calculating the economic and environmental impacts of an emerging discipline is no easy feat. However, evaluating these impacts in the context of one particular application of biofabrication can help us develop a clearer picture of the broader consequences of this technology. Consider, for example, the case study of engineered meat presented in chapter 5. The translation of this research into commercial products is happening now, and it presents a huge economic opportunity. Experts estimate that the global population will total around 10 billion people by the year 2050 and that the consumption of meat by that population will be nearly double that of the world at the turn of the twenty-first century.[1] It is unlikely that we can match the needs of such a massive population using only traditional methods of livestock farming.

Industrial animal farming can be both fiscally and environmentally inefficient. This is because 97% of the calories livestock consume are used purely for maintaining their bodies throughout their lifetimes and producing bodily tissues that humans cannot or will not eat. As a result, much of the energy invested into generating meat, in terms of water and land resources as well as greenhouse gas emissions, is wasted. Growing public concern about the ethical issues surrounding raising animals in high-density settings have resulted in cruelty-free farming techniques. These approaches, while arguably more humane, could increase the costs of meat production by two to three orders of magnitude.[2] Alternative methods must thus be developed in order to meet the rising global demand for cost-effective sources of protein. Recent commercial successes in plant-derived proteins, which could have a smaller environmental impact than animal-sourced foods, have provided one possible alternative to traditional farming. It is unlikely, however, that the cross-cultural global desire for meat products can be fully replaced by plant-based substitutes. Determining whether engineered meat can be produced in a cost-effective and sustainable manner is thus critically important to ensuring worldwide food security.

The initial demonstrations of engineered meat were created in scientific labs as proof-of-concept experiments and, as such, were not designed for affordability or scalability. We can readily see that spending $300,000 on a burger is not a reasonable price point for a common consumable product. Many things that are expensive to produce as one-off goods in a lab are, however, much cheaper to manufacture at scale in a factory setting. Indeed, as we have seen many times since the dawn of the industrial revolution, large-batch purchases of raw goods, automated machinery, optimized process flows, and a skilled workforce can dramatically reduce the costs of building next-generation technologies en masse.

In biotechnology, as in many other industries, there is an inverse relationship between a product's potential market size and its sale price. Bioreactor-based cell culture technologies that have been developed and optimized for other industries, such as microbial fermentation, can be adapted to the production of engineered meat and used to reduce manufacturing costs for producing large numbers of cells. Indeed, it may even make sense to reduce the volume of these bioreactors for the purposes of manufacturing meat, as smaller batch sizes offer greater ability to customize products, adapt to fluctuating market size, and reduce contamination-driven damage.[3] The true financial bottlenecks for manufacturing engineered meat may not, in fact, be the cells themselves or the machines used to grow them but rather the culture media in which they are grown. A recent estimate of the production cost of biofabricated tissues suggests that culture media accounts for 55%–95% of the total cost. Moreover, growth factor proteins account for 99% of the cost of culture media and are thus the predominant limiting factor for reducing manufacturing costs and sale price.[4]

Encouragingly, new methods for manufacturing proteins like growth factors are being developed for a range of medical applications. These technologies are not necessarily designed for large-scale manufacturing, as health care is a low-volume high-margin industry, but adapting them for agriculture may help reduce production costs. Medical products are produced at very high levels of purity that are required for therapeutic applications, but this level of purity is not typically required for generating food-safe nutritional products. Food-safe products can thus generally be produced at much higher volumes with streamlined manufacturing approaches, such as by recycling the media used for multiple batches of engineered tissue. There are already some commercial examples of food-safe protein production. A commercially available plant-based meat mimic, for example, replicates the taste of blood in its products by incorporating engineered heme, the protein responsible for generating this flavor in meat. It is likely that continuing advances in the field of protein manufacturing will reduce the production cost of growth factors used in culture media and, as a result, the future price of engineered meat.

Economic calculations of the feasibility of biofabricated meat are, of course, not the only significant concern underlying adoption of this new technology. Environmental impacts must be predicted and weighed into our collective decision-making on the growth and evolution of this field. An early assessment of the environmental cost of engineering meat estimated that, as compared to traditionally sourced meat, it lowers land use by up to 99%, water use by up to 96%, and energy use by up to 45%. The calculation also revealed that engineered meat had the potential to lower greenhouse gas emissions by up to 96%.[5] This analysis focused on the environmental costs of cell cultivation and growth and not the impact of refrigerated

storage and transport of engineered meat to neighbor-
hoods around the world. As traditional meat farming requires moving whole animals and large volumes of carcass
tissue, however, it is likely that engineering meat with no
excess tissue (i.e., bones, fat, blood) will generate lower
volumes of product and, as a result, reduce environmental
costs for storage and transport.

A deeper environmental calculation performed more
recently delved into aspects of engineering meat beyond
cell culture, such as the cost to make and transport the
microcarrier beads used in bioreactors, the cost of producing serum-free media with appropriate growth factors, and even the cost of cleaning the bioreactors and
managing waste products. This calculation revealed that
engineered meat is likely more energy efficient than traditionally sourced beef but potentially less energy efficient
than traditionally sourced poultry or pork.[6] There is still
significant debate surrounding these calculations, however, as some argue that the climate impact of engineered
meat may only be reduced in the short term. Many of
these environmental cost calculations rely on the metric
of carbon dioxide–equivalent emission, rather than split-
ting emission into the separate categories of carbon dioxide, methane, and nitrous oxide production. As methane
emissions do not accumulate in the same way carbon dioxide emissions do, some researchers argue that the environmental impacts of engineered meat may actually be worse

than traditionally sourced meat over the very long term (i.e., one thousand years).[7] These predictions, however, do not account for new methods of energy production. As we decarbonize our energy sources, the assumptions underlying these calculations will change, impacting our conclusions about the sustainability of biofabricated meat.[8] As such, there is not yet a clear consensus on whether biofabrication will help the environment in the long run, as it cannot be decoupled from ongoing advances in the energy industry. There is still enough evidence to support the possibility that engineered meat is a more sustainable alternative to traditionally sourced meat, however, and this evidence serves as the basis for much ongoing financial investment in this application of biofabrication.

Several companies sponsored by venture capital funding and other sources of investment have generated intellectual property that may enable bringing engineered meat to market. They believe that, while the exact economic value of biofabricated meat has yet to be determined, producing a lot of cells from a small biopsy of a donor animal could yield higher returns per animal than livestock farming. The rise of smaller-scale producers of customized consumable goods, such as those generated by craft breweries and small vineyards for example, also raises the possibility that engineered meat could follow a similar business model. Each producer could generate their own version of a biofabricated meat product by engineering the cells and

There is not yet a clear consensus on whether biofabrication will help the environment in the long run, as it cannot be decoupled from ongoing advances in the energy industry.

surrounding scaffolding in producer-specific ways to generate unique flavors and textures.

While these financial motivations are interesting in theory, there are several practical considerations beyond technical and economic feasibility that must be evaluated in parallel. Companies that grow into viable long-lasting businesses do so because they understand their customers and cater their products to match customer needs and desires. The general population has, on average, positive associations with products such as craft beer and artisanal cheese. These feelings do not, however, readily transfer to customized engineered meat. As discussed in chapter 5, such a high-tech framework for a food product makes people feel uneasy. We already have some evidence of this in the contentious public debate surrounding genetically modified food, for example. Some believe that these perceptions can be changed with good marketing, such as by replacing the term *engineered meat* with the term *cultured meat*. Culturing reminds consumers of fermentation, a process that they are already familiar with for the production of beer and cheese and kombucha, and may potentially increase acceptance.

Beyond consumer acceptance, regulatory pathways for translating these products into consumer goods also remain a significant hurdle. It is possible that engineered meat cannot be legally called meat, as that term has always referred to tissue that grew and matured inside a living animal that was harvested after slaughter. Can tissues produced from an immortalized cell line or derived from a biopsy and grown in a bioreactor be termed meat? Engineered tissues do not form a living creature, though they are made of living cells, and this makes them somewhat undefined objects. This has significant implications on ethics, as discussed in chapter 8, but also has an impact on pathways toward regulatory approval from governmental organizations such as the FDA.

Most FDA guidelines governing the safe production and implementation of donor cells were created to guide medical applications, such as the tissue engineering approaches described in chapter 3. Tissue engineering, however, deals with using human cells that are kept alive and implanted inside the body. By contrast, engineered meat deals with animal cells that are dead prior to ingestion. There are thus many uncertainties regarding whether and how these regulations will change to adapt to this new type of consumer good.

The safe production and implementation of engineered meat is not the only concern of regulatory agencies. Regulation is not just about the safety of biofabricated tissue but also its origin. How can we ensure the cells used to produce biofabricated meat are derived from common meat-producing livestock, such as cows or pigs, and not dangerous animals, endangered species, or even human beings? Such verification is critically necessary and must be addressed before commercial adoption of engineered meat can take place. We must remember, however, that such risks also exist in the production of traditionally sourced meat and are scientifically feasible to mitigate in a cost-effective manner.[9]

The safe, affordable, and sustainable manufacturing of biofabricated meat could have a huge positive impact on our ability to sustain a growing global population with adequate nutrition. Engineered meat may not completely replace traditionally sourced meat, but it will likely supplement such products, as well as plant-derived substitutes, in the future. While cellular agriculture has been used as a case study in this chapter, many of the challenges underlying commercial translation of this technology will be similar to those underlying other real-world applications. Medical applications, for example, will also require producing 3D engineered tissues in carefully regulated commercial facilities rather than research labs. While the types of cells being used and the ways in which they are stored may look quite different, the economic and environmental calculations described in this chapter are still likely to inform how biofabrication is implemented in medicine, robotics, and beyond.

# ETHICAL IMPLICATIONS
# OF BIOFABRICATION

Technological advances do not reside in a moral vacuum. Every real-world implementation of a new technology has implications on the world beyond advancing scientific knowledge, because it impacts society as a whole. The preceding chapters have laid out the case for biofabrication, demonstrating the tremendous potential this discipline has for advancing medicine, agriculture, and beyond. To truly understand any new field, however, it is critical to consider its holistic impact on our broader world. We discussed some of the economic, environmental, and regulatory implications of biofabrication in chapter 7 and have even weighed some of the moral quandaries surrounding human use of animals for medical research and food production in chapters 4 and 5. Mapping out the ethical landscape of building with biology, however, requires looking beyond these topics, which have already been discussed by many others, and asking deeper questions about the definition of life.

Biofabrication involves building an object, such as an organ mimic or a robot, with living cells. This raises the obvious question: is an object made of living materials . . . alive? If there were complete consensus on the definition of life, perhaps this would be an easy question to answer. This is not the case, and, as with many questions at the intersection of basic science and philosophy, there is no single definition of life that is universally accepted. Most definitions have some degree of overlap, however, and the consensus is that a living being is a system that can autonomously metabolize and grow, sense and respond to environmental stimuli, and reproduce.[1] By this description, the biofabricated constructs we have discussed thus far do not fall into the category of living beings. Engineered organs, 3D printed meat, and tissue-powered machines can all certainly metabolize nutrients in surrounding cell culture media and grow in an incubator, but this process requires external support systems and is not autonomous. Moreover, while biofabricated tissues can certainly adapt to biological, chemical, and physical cues from their environment, they cannot reproduce.

The cells that compose biofabricated objects can divide to form new cells, resulting in growth. But this cellular reproduction alone is insufficient to reproduce the overall biofabricated object, as that is composed not only

Mapping out the ethical landscape of building with biology requires asking deeper questions about the definition of life.

of cells but of a range of other materials, such as hydrogels, that cannot self-replicate. Moreover, building the object in a specific 3D architecture and maintaining it in the lab required manufacturing technologies, carefully defined culture media, and laboratory facilities that the object cannot deploy on its own. In other words, a tissue-engineered organ simply cannot autonomously reproduce to create a replica of itself. Thus, it is reasonable to argue that it cannot be considered a living being.

An object not being a living being does not necessarily mean it has no moral status or that it deserves no moral consideration. Consider the example of organ-on-a-chip devices, where biofabricated constructs are engineered with the specific purpose of recapitulating organ function inside the human body. Suppose someone designed a skin-on-a-chip model using animal cells to study how skin senses and responds to touch or temperature. We could use the model to learn how to help people who have been burned by a fire or those who have lost skin sensation due to injury-induced nerve damage. Engineering such a skin-on-a-chip might involve incorporating sensory neurons, a class of cells that can detect pain, into the device. Does this mean that the skin-on-a-chip feels pain? One could reasonably argue that the sensory neuron only *detects* the painful stimulus and does not consciously *feel* pain in the way a human or other living being would. But what if we connected the skin-on-a-chip to other organ-on-a-chip devices, such as a brain-on-a-chip, to create a holistic human-on-a-chip model as discussed in chapter 4? If we engineer a brain tissue mimic that is complex enough, perhaps the pain-sensing neuron in the skin could communicate with other types of neurons in the brain that process pain. One could then make a convincing argument that such a device *feels* pain and that experimentation upon it should be regulated by the same regulatory organizations that govern conducting ethical research on animal subjects.

If this hypothetical organ-on-a-chip device were fabricated with human cells rather than animal cells, the situation could be even more fraught with ethical quandaries.[2] On the plus side, using human cells may reduce the risk and scope of future human clinical trials used to assess the safety and efficacy of new therapies. Moreover, it could enable developing a personalized model of each patient's skin, helping choose the optimal therapy that would work for their bodies. This type of patient-specific therapy, termed *precision medicine*, is one of the fastest growing trends in modern medicine.[3] Personalized experiments, however, despite being focused on the goal of advancing human health, start blurring the line between research performed in a lab and patient care performed in a clinical setting. We are still developing the rules that will help us navigate these murky waters in a safe and ethical manner.

Using human cells in body-on-a-chip devices also raises questions about what kind of consent one would need to obtain from a donor to perform these experiments. Does a human donor own the rights to all future tissues grown from their cells? Several ongoing controversies in biology have already asked this question in different contexts. Henrietta Lacks, the donor whose cancerous tumor cells were used to create a commonly used cell line for medical research (HeLa cells), is a distressing example of a patient whose consent was not obtained prior to cell donation. Her cells have been used to commercialize several medical innovations, but neither she nor her family received any compensation for her part in the research process, and the property rights of human cells remain a hotly contested topic.[4] Beyond financial considerations, it is also concerning that genetic information about Henrietta Lacks's cells is broadly known and distributed in the scientific literature, as this compromises the privacy of her family. This is also a consideration with several commercial genetic testing companies today, which analyze DNA and offer information about ancestry or health. If you choose to share data about your own genes by using these consumer products, you are also inevitably sharing data about your relatives.[5] The ethical implications of this are ill-defined, and we have to work together as a society to cultivate a balanced dialogue about the provenance and ownership of biological materials. These considerations deepen the moral dilemmas facing scientists and doctors as they deploy new biofabricated systems in the real world.

As with most moral dilemmas, there is not always an obvious right answer. Many of us are familiar with the trolley problem, an ethical experiment where one is presented with the option of saving five people being hit by a train by switching the train to a different set of tracks, resulting in the death of one person. There are many variations of this thought experiment, where people are forced to make increasingly uncomfortable choices regarding the value of different types and numbers of human lives.[6] As can be expected, different people make different choices, since they judge the costs and benefits of their decisions based on their own value systems. These value systems are likely informed by their individual experiences, cultures, or religious beliefs and are thus inherently diverse. Making choices involves trade-offs between benefits and costs, and this cost versus benefit analysis is reflected in the moral landscape of nearly every new technology. For example, designing a more environmentally sustainable car could be viewed as the ethical thing to do by those who are concerned about global warming. Making a car more sustainable may involve reducing its weight to decrease its energy consumption, however, and this often comes with the trade-off of making the car less safe in the event of a collision. A car cannot weigh nothing and it cannot be infinitely heavy, but there is also no obvious correct weight for a vehicle. Car companies thus have to make decisions on how safe is safe enough to justify a reduction in energy consumption, and this type of trade-off is common in nearly every industry.[7]

Revisiting the organ-on-a-chip example above with this concept of trade-offs in mind, we can ask: what is the minimum level of complexity we can integrate into an organ-on-a-chip device, such that it generates new medical knowledge without traveling too far into shaky moral territory? Researchers have already started asking these questions, especially in the context of engineering tissues that help us understand the process of human development and birth.[8] Rather than creating a complete model of a human embryo, which would immediately raise ethical questions, researchers often choose to study a specific aspect of birth, such as the development of the amniotic sac. This could help guide clinical care without creating something that has any potential of maturing into a human being. In the same way, in our skin-on-a-chip example, researchers might choose to connect the engineered skin to neurons that sense fluxes in temperature or pressure but explicitly choose not to include neurons that could process that stimulus as pain. It is important to note that these are not current concerns in the field, as we are still very far from being able to engineer tissues that fully replicate the human brain or cognitive function. We may never get there, and that may be because it is too technically difficult or because we choose not to pursue that line of research to avoid ethical complications.

But who should make the decisions on what lines of research should be explored? This question has been asked before, in the context of a range of emerging technologies, from genetic engineering to artificial intelligence. Perhaps the most famous example is the Asilomar Conference in 1975, where biologists, clinicians, and lawyers met to discuss the societal implications of genetically engineered organisms. While the summit generated important guidelines for the biotechnology research community, it failed to include the broader range of stakeholders that would be impacted by this technology. A truly balanced and meaningful discussion would include not only clinicians, but patients. Not only those who interpret the law, but those who vote for laws. Not only those who manufacture the food of tomorrow, but everyone who relies on that food to keep their families alive. This line of thinking has gathered steam, and more experts have started to recognize that a holistic discussion must engage academic researchers, industry professionals, policy makers, journalists, philosophers, and the general public. If we can work together in diverse groups to understand emerging technologies and their impact, it is more likely that we will be able to develop institutional mechanisms to resolve ethical dilemmas.[9]

A holistic discussion can take many forms, and we have yet to form a consensus on how such a diverse group

A truly balanced and meaningful discussion would include not only clinicians, but patients. Not only those who interpret the law, but those who vote for laws. Not only those who manufacture food, but everyone who relies on that food.

of people can productively interact to address moral challenges. A method that I have found particularly useful for discussing the ethical implications of biofabrication is to explore particular scenarios, or vignettes, that push the boundaries of ethics with people of different ages, experiences, and backgrounds.[10] Similar to the skin-on-a-chip device discussed above, this type of exercise helps map out the benefits and risks of pursuing a line of research and define the scope that matches most of the group's moral sensibilities.

Consider, for example, an engineer who biofabricates skeletal muscle to help those who have lost limbs in accidents. What if unpredictable adaptation to its environment leads the muscle to be stronger than natural human muscle? This type of emergent behavior is often discussed in philosophical circles, where a system composed of many individual parts that intercommunicate becomes capable of complex functions that were not engineered but rather evolved or emerged into existence.[11] Unpredictable emergent function poses many ethical questions that the engineer may not have planned for: Would the person who receives the transplant have superhero-like (or supervillain-like) strength? What if people who were never hurt in the first place opted to receive transplants of supermuscle? Would this split the human race into wealthy superstrong individuals and poor normal-strength individuals, perpetuating generational wealth inequality? Is

Cultivating literacy about the ethical considerations of this field will help you shape the institutional mechanisms that define responsible guidelines for researchers.

super-muscle any different than an enhanced prosthetic? How could the engineer have prevented or prepared for this outcome? These questions trigger interesting, exciting, and charged discussions and often reveal that even a group of people who seem homogeneous in age or background can have wildly varying opinions on ethical issues. While a uniform consensus on every question may not be reached, these conversations often help define the lines that most people would not cross, and those lines can be used to establish universal guidelines for researchers.

The moral of this biofabrication story is that scientific innovation is a team sport. Regardless of your background or interests, biofabrication will affect your life. Cultivating literacy about the ethical considerations of this field will, therefore, help you shape the institutional mechanisms that define responsible guidelines for researchers.

# CONCLUSION

The emergence of biofabrication has been swift and, as we have seen throughout the course of this book, has already started to shape the future of a wide range of disciplines. Building on the observation that biological systems have an unparalleled ability to dynamically adapt to their surroundings, scientists and engineers have started to incorporate responsive biological materials into food, medical devices, consumer products, and even robots. These early innovations have shown that building with biology has tremendous potential to advance human health, safety, and productivity.

Building with living materials requires knowing how external signals are processed by individual cells and how groups of cells can work together to generate complex functional behaviors. A rising tide of information from the life sciences community about the structure and function of biological systems laid the foundation upon which biofabrication has been built. Concurrent advances by the engineering community in new manufacturing technologies, such as 3D printing, and new synthetic materials, such as biocompatible hydrogels, provided the tools to accelerate the growth and widespread adoption of biofabrication. We are lucky to bear witness to this timely convergence of biology and engineering, as we will be the first generation to feel the impact that biofabricated systems have on society.

While many of the pioneering advances in biofabrication focused on medical applications, such as tissue engineering and organ-on-a-chip platforms, biofabricated food and consumer products like engineered meat and leather have recently ramped up innovation and translation with admirable rapidity. Certainly, some technical challenges remain before we can encounter biofabricated objects regularly in our daily lives, especially for some of the earlier stage research in areas such as biohybrid robotics. For many biofabricated technologies, however, we have started to transition out of the phase of technical development and into the challenges of real-world implementation. These include defining ownership and intellectual property in the context of biofabricated objects, managing how they will gain regulatory approval, and developing mechanisms for their industrial-scale manufacturing, storage, and transport. These topics require input from experts and stakeholders with a wide range of experience and interests, making now the perfect time for people outside the field of biofabrication to get involved in shaping the future of this discipline.

Real-world implementation of biofabrication will inevitably raise economic, environmental, and ethical questions that require input from diverse stakeholders, including researchers, policy makers, and the general public. It is critical for us, the global community that funds biofabrication research and forms the market for biofabricated products, to cultivate literacy in this field and start shaping the conversation according to our needs and priorities.

For many of us, it will also be important to start thinking about how we can best prepare ourselves and the next generation for the future of work in a biofabricated world. Science and engineering education standards must evolve just as rapidly as the technological advances they enable. This is especially true for interdisciplinary fields, such as biofabrication, where rapid advances in a range of adjacent disciplines can dramatically change the technical landscape.[1] Thankfully, most educators recognize that students must be trained to be lifelong learners, so that they can dynamically integrate new technological advances into the systems they develop. Many core undergraduate classes in science and engineering degree programs, for example, have started integrating project-based learning

It is critical for us, the global community that funds biofabrication research and forms the market for biofabricated products, to cultivate literacy in this field and start shaping the conversation according to our needs and priorities.

approaches that teach students to work in teams to develop solutions to real-world problems.[2] These projects are purposely ill-structured and do not contain carefully defined questions with one correct answer. Rather, they are open-ended and allow for many possible solutions, much like the problems students will encounter in their future careers.

I have had the good fortune to train many young scientists who want to build with biology, both in the research lab and in a classroom setting, and seen living proof that project-based approaches help grow independent and adaptive scientific thinking more effectively than traditional well-defined lab courses.[3] This adaptive thinking does not apply merely to the technical aspects of education but also affects how students identify and address ethical challenges. Early studies have started to establish best practices for integrating bioethics education into college curricula and found that open-ended challenges such as case studies and debates have a significant positive impact on learning outcomes, as compared to traditional lecture-based formats.[4] They help students generate and communicate ideas, consider multiple stakeholder perspectives, and devise a plan that values both technical advancement and community impact.

It is vitally important to recognize that adaptive education in biofabrication and bioethics cannot be confined to college classrooms, nor is it sufficient to only expand the student base to include the realm of K–12 education. People are living longer, and technological landscapes are evolving rapidly. It is no longer acceptable to think of early-life education as the only or best preparation for a career, as this risks leaving older generations behind as the economy changes. Adult retraining programs have already been the subject of public conversation in the context of other emerging technologies, such as automated robots that replaced human beings on manufacturing lines. Too often, however, these conversations happen far too late in the game, after many people have already started losing their jobs and are struggling to transition to new careers. Anticipating these economic changes, and proactively addressing them with programs that continuously adapt to a changing knowledge base, could set biofabrication and other emerging technologies on a more inclusive path moving forward. Thankfully, the biofabrication community has shown itself to be very open to integrating with community maker spaces and hackathons, enabling newcomers of different backgrounds and ages to dive in headfirst and learn through hands-on exposure.

Continuing education in evolving disciplines can also be self-directed. This book serves as an instructive primer to biofabrication in the context of building with living cells derived from mammals. There are many branches of biofabrication that we have touched on but not fully explored in this text that are, nevertheless, of tremendous interest

It is no longer acceptable to think of early-life education as the only or best preparation for a career, as this risks leaving older generations behind as the economy changes.

to the field. For example, there is significant momentum around using genetic engineering tools to change the way living cells naturally behave, and many scientists are interested in building with living cells derived from nonmammalian biological sources, such as plants and insects. These could impact the world in functional areas even beyond those we have discussed, such as architecture, defense, and global security. Luckily, many of the tools and techniques described in this text, as well as the vocabulary we have used to describe different biofabricated objects, are easily translated to these topics. Moreover, the economic, environmental, and ethical challenges we have discussed can readily be adapted to accommodate this broader definition of biofabrication.

The coming decades bring with them some of the biggest challenges we will face as a species. Ten billion human beings will populate the earth by 2050, and we will need sustainable sources of nutrition, safe environments to raise our children, and novel medicines that preserve and promote healthy living. While it is easy to be daunted by this prospect, we must remember that human beings have leveraged scientific innovation as a force for positive societal impact throughout our history. Biofabrication has the potential to address many of the challenges we face but only if we share new knowledge globally and democratize access to information. A technological revolution that positively impacts those in resource-poor as well as

Biofabrication has the potential to address many of the challenges we face but only if we share new knowledge globally and democratize access to information.

resource-rich environments is not only possible but critically necessary.

As we start to build with biology, we may also deepen our appreciation for our planet, which is the sophisticated product of natural forces building with biology for eons. Perhaps we will find that the answers we seek have been hiding in plain sight, in every part of the world around us, all along.

# ACKNOWLEDGMENTS

I am deeply grateful to my mother, Radha Raman, for creating the artwork in this book.

**Adipocytes**
Cells that specialize in the storage of fat

**Biocompatible**
A compound, material, or object that is not harmful to living cells

**Biofabrication**
The act of building an object with biological materials and, in the context of this book, building an object with living cells

**Biohybrid**
An object that is manufactured using both biological materials and synthetic materials

**Bioreactor**
A system that supports the growth, maturation, and maintenance of living cells or engineered tissues through precisely regulated biochemical and mechanical stimulation

**Biosafety Hood**
A cabinet outfitted with a HEPA filter and a fan that circulates sterile air within an internal working area to enable aseptic handling of living cells in a lab

**Bovine**
Derived from or pertaining to cattle

**Cardiomyocytes**
Heart muscle cells

**Collagen**
The main structural protein in the extracellular matrix, often used as a natural hydrogel support system for 3D tissue biofabrication

**Culture Media**
Liquid used to support living cells in culture, generally formed by supplementing water with glucose, salts, vitamins, amino acids, and animal-derived serum

**Endothelial Cells**
The cells that line the inner surfaces of blood in various portions of the body, including blood vessels and organs

**Epithelial Cells**
The cells that line the outer surfaces of various portions of the body, including blood vessels and organs

**Extracellular Matrix**
The 3D network of molecules that surrounds living cells and provides them with structural support and biochemical cues

**Fibrin**
A fibrous protein involved in blood clotting often used as a natural hydrogel for 3D biofabrication of multicellular tissues

**Fibroblasts**
A connective tissue cell that synthesizes extracellular matrix components

**Hepatocytes**
Liver cells

**Hydrogel**
A hydrophilic polymer that is often used as a 3D support structure for living cells in biofabricated tissues

**Immortalized Cell Line**
Cells that were originally derived from an animal but have developed a mutation that lets them proliferate and divide indefinitely in culture media

**Induced Pluripotent Stem Cells**
Stem cells generated by reprogramming mature cells derived from an adult donor

**Mammalian Cell**
A cell derived from a mammal

**Microfluidic Device**
A device that enables manipulating microliter volumes of liquid

**Myoblasts**
Precursors to skeletal muscle cells

**Myotubes**
Contractile muscle fibers formed from the fusion of several myoblasts into an elongated multinucleated structure

**Organoid**
A miniaturized organ mimic derived from stem cells that self-assemble into a 3D structure

**Organ-on-a-Chip**
A microscale organ mimic manufactured by patterning cells in a 3D structure, often cultured within a microfluidic device

**Osteoblasts**
Cells responsible for bone formation

**Primary Cells**
Cells that have been obtained directly from a living animal and thus have many of the important characteristics of the cells in that animal's body but have limited ability to grow and replicate themselves in culture

**Stem Cells**
Cells that have the ability to differentiate into many types of cells inside the body and can grow indefinitely in culture

**3D Printer**
Any of a broad array of machines that perform additive manufacturing of a 3D structure, generally by assembling 2D layers on top of one another

**Tissue Engineering**
Biofabricated multicellular structures that replicate the form and function of tissues or organ systems in the human body

# NOTES

## Chapter 1

1. Joseph A. DiMasi, Henry G. Grabowski, and Ronald W. Hansen, "Innovation in the Pharmaceutical Industry: New Estimates of R&D Costs," *Journal of Health Economics*, no. 47 (2016): 20–33, https://doi.org/10.1016/j.jhealeco.2016.01.012.

## Chapter 2

1. Christina Philippeos, Robin D. Hughes, Anil Dhawan, and Ragai Mitry, "Introduction to Cell Culture," in *Human Cell Culture Protocols*, ed. Ragai R. Mitry and Robin D. Hughes (New York: Springer Science and Business Media, 2012), 1–14, https://doi.org/10.1007/978-1-62703-239-1_1.

2. Tatsuma Yao and Yuta Asayama, "Animal-Cell Culture Media: History, Characteristics, and Current Issues," *Reproductive Medicine and Biology* 16, no. 2 (2017): 99–117, https://doi.org/10.1002/rmb2.12024.

3. Alastair Khodabukus and Keith Baar, "The Effect of Serum Origin on Tissue Engineered Skeletal Muscle Function," *Journal of Cellular Biochemistry* 115, no. 12 (August 2014): 2198–2207, https://doi.org/10.1002/jcb.24938.

4. R. J. Geraghty, A. Capes-Davis, J. M. Davis, J. Downward, R. I. Freshney, I. Knezevic, R. Lovell-Badge et al., "Guidelines for the Use of Cell Lines in Biomedical Research," *British Journal of Cancer* 111, no. 6 (2014): 1021–1046, https://doi.org/10.1038/bjc.2014.166.

5. Gurvinder Kaur and Jannette M. Dufour, "Cell Lines: Valuable Tools or Useless Artifacts," *Spermatogenesis* 2, no. 1 (2012): 1–5, https://doi.org/10.4161/spmg.19885.

6. Keisuke Okita, Tomoko Ichisaka, and Shinya Yamanaka, "Generation of Germline-Competent Induced Pluripotent Stem Cells," *Nature* 448, no. 7151 (2007): 313–317, https://doi.org/10.1038/nature05934.

7. Ranjna C. Dutta and Aroop K. Dutta, *3D Cell Culture* (Singapore: Pan Stanford Publishing, 2018); Francesco Pampaloni, Emmanuel G. Reynaud, and Ernst H. K. Stelzer, "The Third Dimension Bridges the Gap between Cell Culture and Live Tissue," *Nature Reviews* 8, no. 10 (2007): 839–845, https://doi.org/10.1038/nrm2236.

8. Mark W. Tibbitt and Kristi S. Anseth, "Hydrogels as Extracellular Matrix Mimics for 3D Cell Culture," *Biotechnology and Bioengineering* 103, no. 4 (2009): 655–663, https://doi.org/10.1002/bit.22361.

9. Nicholas A. Peppas, J. Zach Hilt, Ali Khademhosseini, and Robert Langer, "Hydrogels in Biology and Medicine: From Molecular Principles to Bionano-technology," *Advanced Materials* 18, no. 11 (2006): 1345–1360, https://doi.org/10.1002/adma.200501612.

10. Yu Shrike Zhang and Ali Khademhosseini, "Advances in Engineering Hy-drogels," *Science* 356, no. 500 (2017): 1–12, https://doi.org/10.1126/science.aaf3627.

11. Piyush Bajaj, Ryan M. Schweller, Ali Khademhosseini, Jennifer L. West, and Rashid Bashir, "3D Biofabrication Strategies for Tissue Engineering and Regenerative Medicine," *Annual Review of Biomedical Engineering* 16, no. 1 (May 2013): 247–276, https://doi.org/10.1146/annurev-bioeng-071813-105155.

12. Sebastien G. M. Uzel, Andrea Pavesi, and Roger D. Kamm, "Microfab-rication and Microfluidics for Muscle Tissue Models," *Progress in Biophysics and Molecular Biology* 115, no. 2–3 (September 2014): 279–293, https://doi.org/10.1016/j.pbiomolbio.2014.08.013.

13. Shauheen S. Soofi, Julie A. Last, Sara J. Liliensiek, Paul F. Nealey, and Christopher J. Murphy, "The Elastic Modulus of Matrigel as Determined by Atomic Force Microscopy," *Journal of Structural Biology* 167, no. 3 (2015): 216–219, https://doi.org/10.1016/j.jsb.2009.05.005.

14. Eduardo Anitua, Paquita Nurden, Roberto Prado, Alan T. Nurden, and Sa-bino Padilla, "Autologous Fibrin Scaffolds: When Platelet- and Plasma-Derived Biomolecules Meet Fibrin," *Biomaterials* 192 (September 2018): 440–460, https://doi.org/10.1016/j.biomaterials.2018.11.029.

15. George M. Whitesides, "The Origins and the Future of Microfluidics," *Nature* 442, no. 7101 (2006): 368–373, https://doi.org/10.1038/nature05058.

16. Helene Andersson and Albert van den Berg, "Microfabrication and Mi-crofluidics for Tissue Engineering: State of the Art and Future Opportunities," *Lab on a Chip* 4, no. 2 (2004): 98–103, https://doi.org/10.1039/b314469k.

17. D. T. Pham and R. S. Gault, "A Comparison of Rapid Prototyping Tech-nologies," *International Journal of Machine Tools and Manufacture* 38, no. 10–11 (1998): 1257–1287, https://doi.org/10.1016/S0890-6955(97)00137-5.

18. Karoly Jakab, Cyrille Norotte, Brook Damon, Francoise Marga, Adrian Neagu, Cynthia L. Besch-Williford, Anatoly Kachurin et al., "Tissue Engi-neering by Self-Assembly of Cells Printed into Topologically Defined Struc-tures," *Tissue Engineering* 14, no. 3 (2008): 413–421, https://doi.org/10.1089/tea.2007.0173; Cyrille Norotte, Francois S. Marga, Laura E. Niklason, and Gabor Forgacs, "Scaffold-Free Vascular Tissue Engineering Using Bioprint-ing," *Biomaterials* 30, no. 30 (2009): 5910–5917, https://doi.org/10.1016/j.biomaterials.2009.06.034.

19. Thomas J. Hinton, Quentin Jallerat, Rachelle N. Palchesko, Joon Hyung Park, Martin S. Grodzicki, Hao-jan Shue, Mohamed H. Ramadan, Andrew R. Hudson, and Adam W. Feinberg, "Three-Dimensional Printing of Complex Biological Structures by Freeform Reversible Embedding of Suspended Hydrogels," *Science Advances* 1, no. 9 (October 2015): 1–10, https://doi.org/10.1126/sciadv.1500758.

20. Esther C. Novosel, Claudia Kleinhans, and Petra J. Kluger, "Vascularization Is the Key Challenge in Tissue Engineering," *Advanced Drug Delivery Reviews* 63, no. 4–5 (2011): 300–311, https://doi.org/10.1016/j.addr.2011.03.004.

21. Mark A. Skylar-Scott, Sebastien G. M. Uzel, Lucy L. Nam, John H. Ahrens, Ryan L. Truby, Sarita Damaraju, and Jennifer A. Lewis, "Biomanufacturing of Organ-Specific Tissues with High Cellular Density and Embedded Vascular Channels," *Science Advances* 5, no. 9 (September 2019).

22. Ritu Raman and Rashid Bashir, "Stereolithographic 3D Bioprinting for Biomedical Applications," in *Essentials of 3D Biofabrication and Translation*, ed. Anthony Atala and James J. Yoo (Waltham, MA: Elsevier, 2015), 89–121.

23. Busaina Dhariwala, Elaine Hunt, and Thomas Boland, "Rapid Prototyping of Tissue-Engineering Constructs, Using Photopolymerizable Hydrogels and Stereolithography," *Tissue Engineering* 10, no. 9/10 (2004): 1316–1322; Karina Arcaute, Brenda K. Mann, and Ryan B. Wicker, "Stereolithography of Three-Dimensional Bioactive Poly(Ethylene Glycol) Constructs with Encapsulated Cells," *Annals of Biomedical Engineering* 34, no. 9 (2006): 1429–1441, https://doi.org/10.1007/s10439-006-9156-y.

24. Benjamin D. Fairbanks, Michael P. Schwartz, Christopher N. Bowman, and Kristi S Anseth, "Photoinitiated Polymerization of PEG-Diacrylate with Lithium Phenul-2,4,6-Trimethylbenzoylphosphinate: Polymerization Rate and Cytocompatibility," *Biomaterials* 30, no. 35 (2009): 6702–6707, https://doi.org/10.1016/j.biomaterials.2009.08.055.

25. Emily R. Ruskowitz and Cole A. DeForest, "Proteome-Wide Analysis of Cellular Response to Ultraviolet Light for Biomaterial Synthesis and Modification," *ACS Biomaterials Science and Engineering* 5, no. 5 (2019), https://doi.org/10.1021/acsbiomaterials.9b00177.

26. Ritu Raman, Nicholas E. Clay, Sanjeet Sen, Molly Melhem, Ellen Qin, Hyunjoon Kong, and Rashid Bashir, "3D Printing Enables Separation of Orthogonal Functions within a Hydrogel Particle," *Biomedical Microdevices* 18, no. 3 (2016): 49, https://doi.org/10.1007/s10544-016-0068-9.

27. Pranav Soman, Jonathan A. Kelber, Jin Woo Lee, Tracy N. Wright, Kenneth S. Vecchio, Richard L. Klemke, and Shaochen Chen, "Cancer Cell Migration within 3D Layer-by-Layer Microfabricated Photocrosslinked PEG

Scaffolds with Tunable Stiffness," *Biomaterials* 33, no. 29 (2012): 7064–7070, https://doi.org/10.1016/j.biomaterials.2012.06.012; Robert Gauvin, Ying-Chieh Chieh Chen, Jin Woo Lee, Pranav Soman, Pinar Zorlutuna, Jason W. Nichol, Hojae Bae, Shaochen Chen, and Ali Khademhosseini, "Microfabrication of Complex Porous Tissue Engineering Scaffolds Using 3D Projection Stereolithography," *Biomaterials* 33, no. 15 (2012): 3824–3834, https://doi.org/10.1016/j.biomaterials.2012.01.048.

28. Ritu Raman, Basanta Bhaduri, Mustafa Mir, Artem Shkumatov, Min Kyung Lee, Gabriel Popescu, Hyunjoon Kong, and Rashid Bashir, "High-Resolution Projection Microstereolithography for Patterning of Neovasculature," *Advanced Healthcare Materials* 5, no. 5 (2015): 610–619, https://doi.org/10.1002/adhm.201500721.

29. R. Daniel Pedde, Bahram Mirani, Ali Navaei, Tara Styan, Sarah Wong, Mehdi Mehrali, Ashish Thakur et al., "Emerging Biofabrication Strategies for Engineering Complex Tissue Constructs," *Advanced Materials* 29, no. 19 (2017): 1–27, https://doi.org/10.1002/ADMA.201606061; Ritu Raman and Rashid Bashir, "Biomimicry, Biofabrication, and Biohybrid Systems: The Emergence and Evolution of Biological Design," *Advanced Healthcare Materials* 6, no. 20 (2017): 1–20, https://doi.org/10.1002/adhm.201700496.

### Chapter 3

1. Linda G. Griffith and Gail Naughton, "Tissue Engineering—Current Challenges and Expanding Opportunities," *Science* 295, no. 5557 (2002): 1009–1014, https://doi.org/10.1126/science.1069210; Ali Khademhosseini and Robert Langer, "A Decade of Progress in Tissue Engineering," *Nature Protocols* 11, no. 10 (2016): 1775–1781, https://doi.org/10.1038/nprot.2016.123.

2. Elsie S. Place, Nicholas D. Evans, and Molly M. Stevens, "Complexity in Biomaterials for Tissue Engineering," *Nature Materials* 8, no. 6 (2009): 457–470, https://doi.org/10.1038/nmat2441.

3. K. E. Healy and R. E. Guldberg, "Bone Tissue Engineering," *Journal of Musculoskeletal & Neuronal Interactions* 7, no. 4 (2007): 328–330, http://www.ncbi.nlm.nih.gov/pubmed/18094496; António J. Salgado, Olga P. Coutinho, and Rui L. Reis, "Bone Tissue Engineering: State of the Art and Future Trends," *Macromolecular Bioscience* 4, no. 8 (2004): 743–765, https://doi.org/10.1002/mabi.200400026; Caleb Vogt, Mitchell Tahtinen, and Feng Zhao, "Engineering Approaches for Creating Skeletal Muscle," *Tissue Engineering and Nanotheranostics* (2017): 1–27, https://doi.org/10.1142/9789813149199_0001; Shannon E. Anderson, Woojin M. Han, Vunya Srinivasa, Mahir Mohiuddin, Marissa A. Ruehle, June Young Moon, Eunjung Shin et al., "Determination

of a Critical Size Threshold for Volumetric Muscle Loss in the Mouse Quadriceps," *Tissue Engineering—Part C: Methods* 25, no. 2 (2019): 59–70, https://doi.org/10.1089/ten.tec.2018.0324.

4. Mohammad R. Ebrahimkhani, Jaclyn A. Shepard Neiman, Micah Sam B. Raredon, David J. Hughes, and Linda G. Griffith, "Bioreactor Technologies to Support Liver Function In Vitro," *Advanced Drug Delivery Reviews* 69 (March 2014): 132–157, https://doi.org/10.1016/j.addr.2014.02.011.

5. Yilin Cao, J. P. Vacanti, Keith T. Paige, Joseph Upton, and Charles A. Vacanti, "Transplantation of Chondrocytes Utilizing a Polymer-Cell Construct to Produce Tissue-Engineered Cartilage in the Shape of a Human Ear," *Plastic and Reconstructive Surgery* 100, no. 2 (1997): 297–302, https://doi.org/10.1097/00006534-199708000-00001.

6. Alyssa J. Reiffel, Concepcion Kafka, Karina A. Hernandez, Samantha Popa, Justin L. Perez, Sherry Zhou, Satadru Pramanik et al., "High-Fidelity Tissue Engineering of Patient-Specific Auricles for Reconstruction of Pediatric Microtia and Other Auricular Deformities," *PloS One* 8, no. 2 (2013): e56506, https://doi.org/10.1371/journal.pone.0056506.

7. Jennifer L. Puetzer and Lawrence J. Bonassar, "Physiologically Distributed Loading Patterns Drive the Formation of Zonally Organized Collagen Structures in Tissue Engineered Meniscus," *Tissue Engineering Part A* 22, no. 607 (2016): 1–40, https://doi.org/10.1089/ten.tea.2015.0519; Stephen R. Sloan, Christoph Wipplinger, Sertaç Kirnaz, Rodrigo Navarro-Ramirez, Franziska Schmidt, Duncan McCloskey, Tania Pannellini, Antonella Schiavinato, Roger Härtl, and Lawrence J. Bonassar, "Combined Nucleus Pulposus Augmentation and Annulus Fibrosus Repair Prevents Acute Intervertebral Disc Degeneration after Discectomy," *Science Translational Medicine* 12, no. 534 (2020): 47–49, https://doi.org/10.1126/scitranslmed.aay2380.

8. Puetzer and Bonassar, "Physiologically Distributed Loading Patterns," 1–40.

9. Nathaniel S. Hwang, Myoung Sook Kim, Somponnat Sampattavanich, Jin Hyen Baek, Zijun Zhang, and Jennifer Elisseeff, "Effects of Three-Dimensional Culture and Growth Factors on the Chondrogenic Differentiation of Murine Embryonic Stem Cells," *Stem Cells* 24, no. 2 (2006): 284–291, https://doi.org/10.1634/stemcells.2005-0024.

10. Gordana Vunjak-Novakovic, Nina Tandon, Amandine Godier, Robert Maidhof, Anna Marsano, Timothy P. Martens, and Milica Radisic, "Challenges in Cardiac Tissue Engineering," *Tissue Engineering* 15, no. 2 (2010): 169–187; Marc N. Hirt, Arne Hansen, and Thomas Eschenhagen, "Cardiac Tissue Engineering: State of the Art," *Circulation Research* 114, no. 2 (2014): 354–367, https://doi.org/10.1161/CIRCRESAHA.114.300522.

11. Jorg Zimmerman, Katharina Bittner, Bjorn Stark, and Rolf Mulhaupt, "Novel Hydrogels as Supports for In Vitro Cell Growth: Poly(Ethylene Glycol)- and Gelatine-Based (Meth)Acrylamidopeptide Macromonomers," *Biomaterials* 23, no. 10 (2002): 2127–2134, http://www.ncbi.nlm.nih.gov/pubmed /11962653.

12. Rebecca L. Carrier, Maria Rupnick, Robert Langer, Frederick J. Schoen, Lisa E. Freed, and Gordana Vunjak-Novakovic, "Perfusion Improves Tissue Architecture of Engineered Cardiac Muscle," *Tissue Engineering* 8, no. 2 (2002): 175–188, https://doi.org/10.1089/107632702753724950.

13. Milica Radisic, Hyoungshin Park, Fen Chen, Johanna E. Salazar-Lazzaro, Yadong Wang, Robert Dennis, Robert Langer, Lisa E. Freed, and Gordana Vunjak-Novakovic, "Biomimetic Approach to Cardiac Tissue Engineering: Oxygen Carriers and Channeled Scaffolds," *Tissue Engineering* 12, no. 8 (2006): 2077–2091, https://doi.org/10.1089/ten.2006.12.2077.

14. Milica Radisic, Hyoungshin Park, Helen Shing, Thomas Consi, Frederick J. Schoen, Robert Langer, Lisa E. Freed, and Gordana Vunjak-Novakovic, "Functional Assembly of Engineered Myocardium by Electrical Stimulation of Cardiac Myocytes Cultured on Scaffolds," *Proceedings of the National Academy of Sciences of the United States of America* 101, no. 52 (2004): 18129–18134, https://doi.org/10.1073/pnas.0407817101.

15. Su Ryon Shin, Sung Mi Jung, Momen Zalabany, Keekyoung Kim, Pinar Zorlutuna, Sang Bok Kim, Mehdi Nikkhah et al., "Carbon-Nanotube-Embedded Hydrogel Sheets for Engineering Cardiac Constructs and Bioactuators," *ACS Nano* 7, no. 3 (2013): 2369–2380, https://doi.org/10.1021/nn305559j.

16. Rohin K. Iyer, Loraine L. Y. Chiu, Lewis A. Reis, and Milica Radisic, "Engineered Cardiac Tissues," *Current Opinion in Biotechnology* 22, no. 5 (2011): 706–714, https://doi.org/10.1016/j.copbio.2011.04.004.

17. Edward L. LeCluyse, Peter L. Bullock, and Andrew Parkinson, "Strategies for Restoration and Maintenance of Normal Hepatic Structure and Function in Long-Term Cultures of Rat Hepatocytes," *Advanced Drug Delivery Reviews* 22, no. 1–2 (1996): 133–186, https://doi.org/10.1016/S0169-409X(96)00418-8; C. Legallais, B. David, and E. Doré, "Bioartificial Livers (BAL): Current Technological Aspects and Future Developments," *Journal of Membrane Science* 181, no. 1 (2001): 81–95, https://doi.org/10.1016/S0376-7388(00)00539-1.

18. Mark J. Powers, Karel Domansky, Mohammad R. Kaazempur-Mofrad, Artemis Kalezi, Adam Capitano, Arpita Upadhyaya, Petra Kurzawski et al., "A Microfabricated Array Bioreactor for Perfused 3D Liver Culture," *Biotechnology and Bioengineering* 78, no. 3 (2001): 257–269, https://doi.org/10.1002 /bit.10143.

19. Jaclyn A. Shepard Neiman, Ritu Raman, Vincent Chan, Mary G. Rhoads, Micha Sam B. Raredon, Jeremy J. Velazquez, Rachel L. Dyer, Rashid Bashir, Paula T. Hammond, and Linda G. Griffith, "Photopatterning of Hydrogel Scaffolds Coupled to Filter Materials Using Stereolithography for Perfused 3D Culture of Hepatocytes," *Biotechnology and Bioengineering* 112, no. 4 (2015): 777–787, https://doi.org/10.1002/bit.25494.

20. Mona Dvir-Ginzberg, Iris Gamlieli-Bonshtein, Riad Agbaria, and Smadar Cohen, "Liver Tissue Engineering within Alginate Scaffolds: Effects of Cell-Seeding Density on Hepatocyte Viability, Morphology, and Function," *Tissue Engineering* 9, no. 4 (2003): 757–766, https://doi.org/10.1089/107632703768247430.

21. Kazuo Ohashi, Takashi Yokoyama, Masayuki Yamato, Hiroyuki Kuge, Hiromichi Kanehiro, Masahiro Tsutsumi, Toshihiro Amanuma et al., "Engineering Functional Two- and Three-Dimensional Liver Systems In Vivo Using Hepatic Tissue Sheets," *Nature Medicine* 13, no. 7 (2007): 880–885, https://doi.org/10.1038/nm1576.

22. Healy and Guldberg, "Bone Tissue Engineering," 328–330; Salgado, Coutinho, and Reis, "Bone Tissue Engineering," 743–765.

23. Adil Al-Mayah, *Biomechanics of Soft Tissues* (Boca Raton, FL: CRC Press Taylor and Francis Group, 2018), 1–27, https://doi.org/10.1017/CBO9781107415324.004.

24. Fiona M. Watt and Wilhelm T. S. Huck, "Role of the Extracellular Matrix in Regulating Stem Cell Fate," *Nature Reviews* 14, no. 8 (2013): 467–473, https://doi.org/10.1038/nrm3620.

25. Lorenz Meinel, Vassilis Karageorgiou, Robert Fajardo, Brian Snyder, Vivek Shinde-Patil, Ludwig Zichner, David Kaplan, Robert Langer, and Gordana Vunjak-Novakovic, "Bone Tissue Engineering Using Human Mesenchymal Stem Cells: Effects of Scaffold Material and Medium Flow," *Annals of Biomedical Engineering* 32, no. 1 (2004): 112–122, https://doi.org/10.1023/B:ABME.0000007796.48329.b4.

26. A. Ignatius, H. Blessing, A. Liedert, C. Schmidt, C. Neidlinger-Wilke, D. Kaspar, B. Friemert, and L. Claes, "Tissue Engineering of Bone: Effects of Mechanical Strain on Osteoblastic Cells in Type I Collagen Matrices," *Biomaterials* 26, no. 3 (2005): 311–318, https://doi.org/10.1016/j.biomaterials.2004.02.045.

27. Z. Schwartz, B. J. Simon, M. A. Duran, G. Barabino, R. Chaudhri, and B. D. Boyan, "Pulsed Electromagnetic Fields Enhance BMP-2 Dependent Osteoblastic Differentiation of Human Mesenchymal Stem Cells," *Journal of Orthopaedic Research* 26, no. 9 (2008): 1250–1255, https://doi.org/10.1002/jor.20591.

28. Jordan S. Miller, Kelly R. Stevens, Michael T. Yang, Brendon M. Baker, Duc-Huy T. Nguyen, Daniel M. Cohen, Esteban Toro et al., "Rapid Casting of Patterned Vascular Networks for Perfusable Engineered Three-Dimensional Tissues," *Nature Materials* 11, no. 9 (2012): 768–774, https://doi.org/10.1038/nmat3357.

29. April M. Kloxin, Andrea M. Kasko, Chelsea N. Salinas, and Kristi S. Anseth, "Photodegradable Hydrogels for Dynamic Tuning of Physical and Chemical Properties," *Science* 324, no. 5923 (2009): 59–63, https://doi.org/10.1126/science.1169494; Cole A. DeForest and Kristi S. Anseth, "Cytocompatible Click-Based Hydrogels with Dynamically Tunable Properties through Orthogonal Photoconjugation and Photocleavage Reactions," *Nature Chemistry* 3, no. 12 (2011): 925–931, https://doi.org/10.1038/NCHEM.1174; Ritu Raman, Tiffany Hua, Declan Gwynne, Joy Collins, Siid Tamang, Vance Soares, Tina Esfandiary et al., "Light-Degradable Hydrogels as Dynamic Triggers in Gastrointestinal Applications," *Science Advances* 6, no. 3 (January 2020): 1–12.

30. Christopher K. Arakawa, Barry A. Badeau, Ying Zheng, and Cole A. DeForest, "Multicellular Vascularized Engineered Tissues through User-Programmable Biomaterial Photodegradation," *Advanced Materials* 29, no. 37 (2017): 1703156, https://doi.org/10.1002/adma.201703156.

31. Ritu Raman, Basanta Bhaduri, Mustafa Mir, Artem Shkumatov, Min Kyung Lee, Gabriel Popescu, Hyunjoon Kong, and Rashid Bashir, "High-Resolution Projection Microstereolithography for Patterning of Neovasculature," *Advanced Healthcare Materials* 5, no. 5 (2015): 610–619, https://doi.org/10.1002/adhm.201500721.

32. Molly R. Melhem, Jooyeon Park, Luke Knapp, Larissa Reinkensmeyer, Caroline Cvetkovic, Jordan Flewellyn, Min Kyung Lee et al., "3D Printed Stem-Cell-Laden, Microchanneled Hydrogel Patch for the Enhanced Release of Cell-Secreting Factors and Treatment of Myocardial Infarctions," *ACS Biomaterials Science and Engineering* 3, no. 9 (2017): 1980–1987, https://doi.org/10.1021/acsbiomaterials.6b00176.

33. Masashi Nomi, Anthony Atala, Paolo De Coppi, and Shay Soker, "Principals of Neovascularization for Tissue Engineering," *Molecular Aspects of Medicine* 23, no. 6 (2002): 463–483, https://doi.org/10.1016/S0098-2997(02)00008-0; Richard P. Visconti, Vladimir Kasyanov, Carmine Gentile, Jing Zhang, Roger R. Markwald, and Vladimir Mironov, "Towards Organ Printing: Engineering an Intra-Organ Branched Vascular Tree," *Expert Opinion on Biological Therapy* 10, no. 3 (2010): 409–420, https://doi.org/10.1517/14712590903563352.

34. Jeremy J. Song, Jacques P. Guyette, Sarah E. Gilpin, Gabriel Gonzalez, Joseph P. Vacanti, and Harald C. Ott, "Regeneration and Experimental Orthotopic

Transplantation of a Bioengineered Kidney," *Nature Medicine* 19, no. 5 (2013): 646–651, https://doi.org/10.1038/nm.3154; Midori Kato-Negishi, Yuya Morimoto, Hiroaki Onoe, and Shoji Takeuchi, "Millimeter-Sized Neural Building Blocks for 3D Heterogeneous Neural Network Assembly," *Advanced Healthcare Materials* 2, no. 12 (2013): 1564–1570, https://doi.org/10.1002/adhm.20130 0052; Justine R. Yu, Javier Navarro, James C. Coburn, Bhushan Mahadik, Joseph Molnar, James H. Holmes, Arthur J. Nam, and John P. Fisher, "Current and Future Perspectives on Skin Tissue Engineering: Key Features of Biomedical Research, Translational Assessment, and Clinical Application," *Advanced Healthcare Materials* 8, no. 5 (2019): 1–19, https://doi.org/10.1002/adhm .201801471.

## Chapter 4

1.  Joseph A. DiMasi, Henry G. Grabowski, and Ronald W. Hansen, "Innovation in the Pharmaceutical Industry: New Estimates of R&D Costs," *Journal of Health Economics* 47 (2016): 20–33, https://doi.org/10.1016/j.jhealeco .2016.01.012.

2.  Tinneke Denayer, Thomas Stöhrn, and Maarten Van Roy, "Animal Models in Translational Medicine: Validation and Prediction," *New Horizons in Translational Medicine* 2, no. 1 (2014): 5–11, https://doi.org/10.1016/j.nhtm .2014.08.001.

3.  Ranjna C. Dutta and Aroop K. Dutta, *3D Cell Culture* (Singapore: Pan Stanford Publishing, 2018).

4.  Neil Kaplowitz, "Idiosyncratic Drug Hepatotoxicity," *Nature Reviews Drug Discovery* 4, no. 6 (2005): 489–499, https://doi.org/10.1038/nrd1750.

5.  Salman R. Khetani and Sangeeta N. Bhatia, "Microscale Culture of Human Liver Cells for Drug Development," *Nature Biotechnology* 26, no. 1 (2008): 120–126, https://doi.org/10.1038/nbt1361.

6.  Sangeeta N. Bhatia and Donald E. Ingber, "Microfluidic Organs-on-Chips," *Nature Biotechnology* 32, no. 8 (2014): 760–772, https://doi.org/10.1038/nbt .2989.

7.  Mahmut Selman Sakar, Jeroen Eyckmans, Roel Pieters, Daniel Eberli, Bradley J. Nelson, and Christopher S. Chen, "Cellular Forces and Matrix Assembly Coordinate Fibrous Tissue Repair," *Nature Communications* 7, no. 1 (2016): 1–8, https://doi.org/10.1038/ncomms11036.

8.  Serge Ostrovidov, Vahid Hosseini, Samad Ahadian, Toshinori Fujie, Selvakumar Prakash Parthiban, Murugan Ramalingam, Hojae Bae, Hirokazu Kaji, and Ali Khademhosseini, "Skeletal Muscle Tissue Engineering: Methods to Form

Skeletal Myotubes and Their Applications," *Tissue Engineering Part B* 20, no. 5 (2014): 403–436, https://doi.org/10.1089/ten.teb.2013.0534; Giorgio Cittadella Vigodarzere and Sara Mantero, "Skeletal Muscle Tissue Engineering: Strategies for Volumetric Constructs," *Frontiers in Physiology* 5 (2014): 362, https://doi.org/10.3389/fphys.2014.00362.

9.  Weining Bian and Nenad Bursac, "Engineered Skeletal Muscle Tissue Networks with Controllable Architecture," *Biomaterials* 30, no. 7 (2009): 1401–1412, https://doi.org/10.1016/j.biomaterials.2008.11.015; Weining Bian, Brian Liau, Nima Badie, and Nenad Bursac, "Mesoscopic Hydrogel Molding to Control the 3D Geometry of Bioartificial Muscle Tissues," *Nature Protocols* 4, no. 10 (2009): 1522–1534, https://doi.org/10.1038/nprot.2009.155.

10.  Herman Vandenburgh, Janet Shansky, Frank Benesch-Lee, Kirsten Skelly, Janelle M. Spinazzola, Yero Saponjian, and Brian S. Tseng, "Automated Drug Screening with Contractile Muscle Tissue Engineered from Dystrophic Myoblasts," *FASEB Journal* 23, no. 10 (2009): 3325–3334, https://doi.org/10.1096/fj.09-134411.

11.  Erik Landfeldt, Rachel Thompson, Thomas Sejersen, Hugh J. McMillan, Janbernd Kirschner, and Hanns Lochmüller, "Life Expectancy at Birth in Duchenne Muscular Dystrophy: A Systematic Review and Meta-Analysis," *European Journal of Epidemiology* 35, no. 0123456789 (2020): 1–11, https://doi.org/10.1007/s10654-020-00613-8.

12.  Sebastien G. M. Uzel, Randall J. Platt, Vidya Subramanian, Taylor M. Pearl, Christopher J. Rowlands, Vincent Chan, Laurie A. Boyer, Peter T. C. So, and Roger D Kamm, "Microfluidic Platform for the Formation of Optically Excitable, Three-Dimensional, Compartmentalized Motor Units," *Science Advances* 2, no. 8 (August 2016): 1–13.

13.  Tatsuya Osaki, Sebastien G. M. Uzel, and Roger D. Kamm, "Microphysiological 3D Model of Amyotrophic Lateral Sclerosis (ALS) from Human IPS-Derived Muscle Cells and Optogenetic Motor Neurons," *Science Advances* 4, no. 10 (2018): eaat5847, https://doi.org/10.1126/sciadv.aat5847.

14.  Tatsuya Osaki, Sebastien G. M. Uzel, and Roger D. Kamm, "On-Chip 3D Neuromuscular Model for Drug Screening and Precision Medicine in Neuromuscular Disease," *Nature Protocols* 15, no. 2 (2020): 421–449, https://doi.org/10.1038/s41596-019-0248-1.

15.  Dongeun Huh, Benjamin D. Matthews, Akiko Mammoto, Martín Montoya-Zavala, Hong Yuan Hsin, and Donald E. Ingber, "Reconstituting Organ-Level Lung Functions on a Chip," *Science* 328, no. 5986 (2010): 1662–1668, https://doi.org/10.1126/science.1188302.

16. Dorotaeuvenbe Napierska, Leen C. J. Thomassen, Virginie Rabolli, Dominique Lison, Laetitia Gonzalez, Micheline Kirsch-Volders, Johan A. Martens, and Peter H. Hoet, "Size-Dependent Cytotoxicity of Monodisperse Silica Nanoparticles in Human Endothelial Cells," *Small* 5, no. 7 (2009): 846–853, https://doi.org/10.1002/smll.200800461.

17. Hyun Jung Kim, Dongeun Huh, Geraldine Hamilton, and Donald E. Ingber, "Human Gut-on-a-Chip Inhabited by Microbial Flora That Experiences Intestinal Peristalsis-like Motions and Flow," *Lab on a Chip* 12, no. 12 (2012): 2165–2174, https://doi.org/10.1039/c2lc40074j.

18. Yoojin Shin, Se Hoon Choi, Eunhee Kim, Enjana Bylykbashi, Jeong Ah Kim, Seok Chung, Doo Yeon Kim, Roger D. Kamm, and Rudolph E. Tanzi, "Blood–Brain Barrier Dysfunction in a 3D In Vitro Model of Alzheimer's Disease," *Advanced Science* 6, no. 2 (2019): 1900962, https://doi.org/10.1002/advs.201900962.

19. Anna Herland, Andries D. Van Der Meer, Edward A. FitzGerald, Tae Eun Park, Jelle J. F. Sleeboom, and Donald E. Ingber, "Distinct Contributions of Astrocytes and Pericytes to Neuroinflammation Identified in a 3D Human Blood-Brain Barrier on a Chip," *PLoS ONE* 11, no. 3 (2016): 1–22, https://doi.org/10.1371/journal.pone.0150360.

20. Boyang Zhang, Miles Montgomery, M. Dean Chamberlain, Shinichiro Ogawa, Anastasia Korolj, Aric Pahnke, Laura A. Wells et al., "Biodegradable Scaffold with Built-in Vasculature for Organ-on-a-Chip Engineering and Direct Surgical Anastomosis," *Nature Materials* 15, no. 6 (2016): 669–678, https://doi.org/10.1038/nmat4570; Peter Loskill, Thiagarajan Sezhian, Kevin M. Tharp, Felipe T. Lee-Montiel, Shaheen Jeeawoody, Willie Mae Reese, Peter James H. Zushin, Andreas Stahl, and Kevin E. Healy, "WAT-on-a-Chip: A Physiologically Relevant Microfluidic System Incorporating White Adipose Tissue," *Lab on a Chip* 17, no. 9 (2017): 1645–1654, https://doi.org/10.1039/c6lc01590e; Maierdanjiang Wufuer, Geon Hui Lee, Woojune Hur, Byoungjun Jeon, Byung Jun Kim, Tae Hyun Choi, and Sang Hoon Lee, "Skin-on-a-Chip Model Simulating Inflammation, Edema and Drug-Based Treatment," *Scientific Reports* 6 (October 2016): 1–12, https://doi.org/10.1038/srep37471; Jeongyun Seo, Woo Y. Byun, Farid Alisafaei, Andrei Georgescu, Yoon Suk Yi, Mina Massaro-Giordano, Vivek B. Shenoy, Vivian Lee, Vatinee Y. Bunya, and Dongeun Huh, "Multiscale Reverse Engineering of the Human Ocular Surface," *Nature Medicine* 25, no. 8 (2019): 1310–1318, https://doi.org/10.1038/s41591-019-0531-2; Sijie Hao, Laura Ha, Gong Cheng, Yuan Wan, Yiqiu Xia, Donna M. Sosnoski, Andrea M. Mastro, and Si Yang Zheng, "A Spontaneous

3D Bone-On-a-Chip for Bone Metastasis Study of Breast Cancer Cells," *Small* 14, no. 12 (2018): 1–10, https://doi.org/10.1002/smll.201702787.

21. Ben M. Maoz, Anna Herland, Edward A. FitzGerald, Thomas Grevesse, Charles Vidoudez, Alan R. Pacheco, Sean P. Sheehy et al., "A Linked Organ-on-Chip Model of the Human Neurovascular Unit Reveals the Metabolic Coupling of Endothelial and Neuronal Cells," *Nature Biotechnology* 36, no. 9 (April 2017): 865–874, https://doi.org/10.1038/nbt.4226.

22. Hasan Erbil Abaci and Michael L. Shuler, "Human-on-a-Chip Design Strategies and Principles for Physiologically Based Pharmocokinetics/Pharmacodynamics Modeling Hasan," *Integrative Biology* 7, no. 4 (2015): 383–391, https://doi.org/10.1016/j.physbeh.2017.03.040.

23. Ritu Raman, Erin B. Rousseau, Michael Wade, Allison Tong, Max J. Cotler, Jenevieve Kuang, Alejandro Aponte Lugo et al., "Platform for Micro-Invasive Membrane-Free Biochemical Sampling of Brain Interstitial Fluid," *Science Advances* 6, no. 39 (2020): eabb0657, https://advances.sciencemag.org/content/6/39/eabb0657.

24. Waseem K. Raja, Alison E. Mungenast, Yuan Ta Lin, Tak Ko, Fatema Abdur-rob, Jinsoo Seo, and Li Huei Tsai, "Self-Organizing 3D Human Neural Tissue Derived from Induced Pluripotent Stem Cells Recapitulate Alzheimer's Disease Phenotypes," *PLoS ONE* 11, no. 9 (2016): 1–18, https://doi.org/10.1371/journal.pone.0161969.

25. Sunghee Estelle Park, Andrei Georgescu, and Dongeun Huh, "Organoids-on-a-Chip," *Science* 364, no. 6444 (June 2019): 960–965.

**Chapter 5**

1. P. D. Edelman, D. C. McFarland, V. A. Mironov, and J. G. Matheny, "In Vitro-Cultured Meat Production," *Tissue Engineering* 11, no. 5–6 (2005): 659–662; Hanna L. Tuomisto and M. Joost Teixeira De Mattos, "Environmental Impacts of Cultured Meat Production," *Environmental Science and Technology* 45, no. 14 (2011): 6117–6123, https://doi.org/10.1021/es200130u.

2. Cor van der Weele and Johannes Tramper, "Cultured Meat: Every Village Its Own Factory?," *Trends in Biotechnology* 32, no. 6 (2014): 294–296, https://doi.org/10.1016/j.tibtech.2014.04.009.

3. M. A. Benjaminson, J. A. Gilchriest, and M. Lorenz, "In Vitro Edible Muscle Protein Production System (MPPS): Stage 1, Fish," *Acta Astronautica* 51, no. 12 (2002): 879–889, https://doi.org/10.1016/S0094-5765(02)00033-4.

4. A. Listrat, B. Lebret, I. Louveau, T. Astruc, M. Bonnet, L. Lefaucheur, and J. Bugeon, "How Muscle Structure and Composition Determine Meat Quality," *Productions Animales* 28, no. 2 (2015): 125–136.

5. Alina Surmacka Szczesniak, "Texture Is a Sensory Property," *Food Quality and Preference* 13, no. 4 (2002): 215–225, https://doi.org/10.1016/S0950 -3293(01)00039-8.

6. van der Weele and Tramper, "Cultured Meat," 294–296.

7. Mark J. Post, "Cultured Meat from Stem Cells: Challenges and Prospects," *Meat Science* 92, no. 3 (2012): 297–301, https://doi.org/10.1016/j .meatsci.2012.04.008.

8. Vincent Bodiou, Panagiota Moutsatsou, and Mark J. Post, "Microcarriers for Upscaling Cultured Meat Production," *Frontiers in Nutrition* 7 (February 2020): 1–16, https://doi.org/10.3389/fnut.2020.00010.

9. Scott J. Allan, Paul A. De Bank, and Marianne J. Ellis, "Bioprocess Design Considerations for Cultured Meat Production with a Focus on the Expansion Bioreactor," *Frontiers in Sustainable Food Systems* 3, no. 44 (June 2019): 1–9, https://doi.org/10.3389/fsufs.2019.00044.

10. Sanne Verbruggen, Daan Luining, Anon van Essen, and Mark J. Post, "Bovine Myoblast Cell Production in a Microcarriers-Based System," *Cytotechnology* 70, no. 2 (2018): 503–512, https://doi.org/10.1007/s10616-017-0101-8.

11. Bodiou, Moutsatsou, and Post, "Microcarriers," 1–16.

12. Luke A. Macqueen, Charles G. Alver, Christophe O. Chantre, Seungkuk Ahn, Luca Cera, Grant M. Gonzalez, Blakely B. O. Connor et al., "Muscle Tissue Engineering in Fibrous Gelatin: Implications for Meat Analogs," *Npj Science of Food* 3, no. 20 (February 2019): 1–12, https://doi.org/10.1038 /s41538-019-0054-8.

13. Tom Ben-Arye, Yulia Shandalov, Shahar Ben-Shaul, Shira Landau, Yedidya Zagury, Iris Ianovici, Neta Lavon, and Shulamit Levenberg, "Textured Soy Protein Scaffolds Enable the Generation of Three-Dimensional Bovine Skeletal Muscle Tissue for Cell-Based Meat," *Nature Food* 1, no. 4 (2020): 210–220.

14. Kangmin Seo, Takahiro Suzuki, Ken Kobayashi, and Takanori Nishimura, "Adipocytes Suppress Differentiation of Muscle Cells in a Co-Culture System," *Animal Science Journal* 90, no. 3 (2019): 423–434, https://doi.org/10.1111 /asj.13145.

15. Natalie R. Rubio, Kyle D. Fish, Barry A. Trimmer, and David L. Kaplan, "Possibilities for Engineered Insect Tissue as a Food Source," *Frontiers in Sustainable Food Systems* 3, no. 24 (April 2019): 1–13, https://doi.org/10.3389 /fsufs.2019.00024.

16. Christopher Bryant and Courtney Dillard, "The Impact of Framing on Acceptance of Cultured Meat," *Frontiers in Nutrition* 6, no. 3 (July 2019): 1–10, https://doi.org/10.3389/fnut.2019.00103.

17. Evan Hertafeld, Connie Zhang, Zeyuan Jin, Abigail Jakub, Katherine Russell, Yadir Lakehal, Kristina Andreyeva et al., "Multi-Material Three-Dimensional Food Printing with Simultaneous Infrared Cooking," *3D Printing and Additive Manufacturing* 6, no. 1 (2019): 13–19, https://doi.org/10.1089/3dp.2018.0042.

18. K. Jakab, F. Marga, R. Kaesser, T.-H. Chuang, H. Varadaraju, D. Cassingham, S. Lee, A. Forgacs, and G. Forgacs, "Non-Medical Applications of Tissue Engineering: Biofabrication of a Leather-like Material," *Materials Today Sustainability* 5 (2019): 100018, https://doi.org/10.1016/j.mtsust.2019.100018.

19. Evelyne Battaglia Richi, Beatrice Baumer, Beatrice Conrad, Roger Darioli, Alexandra Schmid, and Ulrich Keller, "Health Risks Associated with Meat Consumption: A Review of Epidemiological Studies," *International Journal for Vitamin and Nutrition Research* 85, no. 1–2 (2015): 70–78, https://doi.org/10.1024/0300-9831/a000224.

20. Niall Firth, "Cultured Meat Has Been Approved for Consumers for the First Time," *MIT Technology Review*, December 1, 2020, https://www.technologyreview.com/2020/12/01/1012789/cultured-cultivated-meat-just-singapore-approved-food-climate/.

**Chapter 6**

1. Leonardo Ricotti, Barry Trimmer, Adam W. Feinberg, Ritu Raman, Kevin K. Parker, Rashid Bashir, Metin Sitti, Sylvain Martel, Paolo Dario, and Arianna Menciassi, "Biohybrid Actuators for Robotics: A Review of Devices Actuated by Living Cells," *Science Robotics* 2, no. 12 (2017): 1–18, https://doi.org/10.1126/scirobotics.aaq0495; Victoria A. Webster-Wood, Ozan Akkus, Umut A. Gurkan, Hillel J. Chiel, and Roger D. Quinn, "Organismal Engineering: Toward a Robotic Taxonomic Key for Devices Using Organic Materials," *Science Robotics* 2, no. 12 (2017): 1–19; Chuang Zhang, Wenxue Wang, Ning Xi, Yuechao Wang, and Lianqing Liu, "Development and Future Challenges of Bio-Syncretic Robots," *Engineering* 4, no. 4 (2018): 452–463, https://doi.org/10.1016/j.eng.2018.07.005.

2. Ritu Raman, "Modeling Muscle," *Science* 363, no. 6431 (2019): 1051; Rebecca M. Duffy and Adam W. Feinberg, "Engineered Skeletal Muscle Tissue for Soft Robotics: Fabrication Strategies, Current Applications, and Future Challenges," *Wiley Interdisciplinary Reviews: Nanomedicine and Nanobiotechnology* 6, no. 2 (2014): 178–195, https://doi.org/10.1002/wnan.1254; Devin Michael Neal, Mahmut Selman Sakar, and H. Harry Asada, "Optogenetic Control of Live Skeletal Muscles: Non-Invasive, Wireless, and Precise Activation

of Muscle Tissues," in *American Control Conference* (Washington, DC: IEEE, 2013), 1513–1518.

3. Hugh Herr and Robert G. Dennis, "A Swimming Robot Actuated by Living Muscle Tissue," *Journal of Neuroengineering and Rehabilitation* 1, no. 1 (2004): 6, https://doi.org/10.1186/1743-0003-1-6.

4. Adam W. Feinberg, Alex Feigel, Sergey S. Shevkoplyas, Sean Sheehy, George M. Whitesides, and Kevin Kit Parker, "Muscular Thin Films for Building Actuators and Powering Devices," *Science* 317, no. 5843 (2007): 1366–1370, https://doi.org/10.1126/science.1146885.

5. Janna C. Nawroth, Hyungsuk Lee, Adam W. Feinberg, Crystal M. Ripplinger, Megan L. McCain, Anna Grosberg, John O. Dabiri, and Kevin Kit Parker, "A Tissue-Engineered Jellyfish with Biomimetic Propulsion," *Nature Biotechnology* 30, no. 8 (2012): 792–797, https://doi.org/10.1038/nbt.2269.

6. Brian J. Williams, Sandeep V. Anand, Jagannathan Rajagopalan, and M. Taher A. Saif, "A Self-Propelled Biohybrid Swimmer at Low Reynolds Number," *Nature Communications* 5, no. 1 (January 2014): 1–8, https://doi.org/10.1038/ncomms4081.

7. Su Ryon Shin, Sung Mi Jung, Momen Zalabany, Keekyoung Kim, Pinar Zorlutuna, Sang Bok Kim, Mehdi Nikkhah et al., "Carbon-Nanotube-Embedded Hydrogel Sheets for Engineering Cardiac Constructs and Bioactuators," *ACS Nano* 7, no. 3 (2013): 2369–2380, https://doi.org/10.1021/nn305559j.

8. Caroline Cvetkovic, Ritu Raman, Vincent Chan, Brian J. Williams, Madeline Tolish, Piyush Bajaj, Mahmut Selman Sakar, H. Harry Asada, M. Taher A. Saif, and Rashid Bashir, "Three-Dimensionally Printed Biological Machines Powered by Skeletal Muscle," *Proceedings of the National Academy of Sciences of the United States of America* 111, no. 28 (2014): 10125–10130, https://doi.org/10.1073/pnas.1401577111.

9. Ritu Raman, Caroline Cvetkovic, Sebastien G. M. Uzel, Randall J. Platt, Parijat Sengupta, and Roger D. Kamm, "Optogenetic Skeletal Muscle-Powered Adaptive Biological Machines," *Proceedings of the National Academy of Sciences* 113, no. 13 (2016): 3497–3502, https://doi.org/10.1073/pnas.1516139113.

10. Zhengwei Li, Yongbeom Seo, Onur Aydin, Mohamed Elhebeary, Roger D. Kamm, Hyunjoon Kong, and M. Taher A. Saif, "Biohybrid Valveless Pump-Bot Powered by Engineered Skeletal Muscle," *Proceedings of the National Academy of Sciences* 116, no. 5 (2019): 1543–1548, https://www.pnas.org/content/116/5/1543.short.

11. S.-J. Park, M. Gazzola, K. S. Park, S. Park, V. Di Santo, E. L. Blevins, J. U. Lind et al., "Phototactic Guidance of a Tissue-Engineered Soft-Robotic Ray,"

*Science* 353, no. 6295 (2016): 158–162, https://doi.org/10.1126/science.aaf 4292.

12. Yoshitake Akiyama, Toru Sakuma, Kei Funakoshi, Takayuki Hoshino, Kikuo Iwabuchi, and Keisuke Morishima, "Atmospheric-Operable Bioactuator Powered by Insect Muscle Packaged with Medium," *Lab on a Chip* 13, no. 24 (2013): 4870–4880, https://doi.org/10.1039/C3LC50490E; Eitaro Yamatsuta, Sze Ping Beh, Kaoru Uesugi, Hidenobu Tsujimura, and Keisuke Morishima, "A Micro Peristaltic Pump Using an Optically Controllable Bioactuator," *Engineering* 5, no. 3 (2019): 580–585, https://doi.org/10.1016/j.eng.2018.11.033; Victoria A. Webster, Emma L. Hawley, Ozan Akkus, Hillel J. Chiel, and Roger D. Quinn, "Effect of Actuating Cell Source on Locomotion of Organic Living Machines with Electrocompacted Collagen Skeleton," *Bioinspiration and Biomimetics* 11, no. 3 (2016): 036012, https://doi.org/10.1088/1748-3190 /11/3/036012; Rika Wright Carlsen, Matthew R. Edwards, Jiang Zhuang, Cecile Pacoret, and Metin Sitti, "Magnetic Steering Control of Multi-Cellular Bio-Hybrid Microswimmers," *Lab on a Chip* 14, no. 19 (2014): 3850–3859, https://doi.org/10.1039/c4lc00707g; Jiang Zhuang, Rika Wright Carlsen, and Metin Sitti, "PH-Taxis of Biohybrid Microsystems," *Scientific Reports* 5 (2015): 11403, https://doi.org/10.1038/srep11403; Clement Appiah, Christine Arndt, Katharina Siemsen, Anne Heitmann, Anne Staubitz, and Christine Selhuber-Unkel, "Living Materials Herald a New Era in Soft Robotics," *Advanced Materials* 31, no. 36 (2019): 1807747, https://doi.org/10.1002/adma.201807747.

13. Swathi Rangarajan, Lauran Madden, and Nenad Bursac, "Use of Flow, Electrical, and Mechanical Stimulation to Promote Engineering of Striated Muscles," *Annals of Biomedical Engineering* 42, no. 7 (2014): 1391–1405, https://doi.org/10.1007/s10439-013-0966-4.

14. Raman et al., "Optogenetic Skeletal Muscle-Powered Adaptive Biological Machines."

15. Ritu Raman, Lauren Grant, Yongbeom Seo, Caroline Cvetkovic, Michael Gapinske, Alexandra Palasz, Howard Dabbous, Hyunjoon Kong, Pablo Perez Pinera, and Rashid Bashir, "Damage, Healing, and Remodeling in Optogenetic Skeletal Muscle Bioactuators," *Advanced Healthcare Materials* 6, no. 12 (2017): 1700030, https://doi.org/10.1002/adhm.201700030.

16. Roger D. Kamm, Rashid Bashir, Natasha Arora, Roy D. Dar, Martha U. Gillette, Linda G. Griffith, Melissa L. Kemp et al., "Perspective: The Promise of Multi-Cellular Engineered Living Systems," *APL Bioengineering* 2, no. 4 (2018): 040901, https://doi.org/10.1063/1.5038337; Caroline Cvetkovic, Max H. Rich, Ritu Raman, Hyunjoon Kong, and Rashid Bashir, "A 3D-Printed Platform

for Modular Neuromuscular Motor Units," *Microsystems and Nanoengineering* 3, no. 1 (2017): 1–9, https://doi.org/10.1038/micronano.2017.15; Onur Aydin, Xiaotian Zhang, Sittinon Nuethong, Gelson J. Pagan-Diaz, Rashid Bashir, Mattia Gazzola, and M. Taher A. Saif, "Neuromuscular Actuation of Biohybrid Motile Bots," *Proceedings of the National Academy of Sciences* 116, no. 40 (2019): 19841–19847, https://doi.org/10.1073/pnas.1907051116; Ritu Raman, Caroline Cvetkovic, and Rashid Bashir, "A Modular Approach to the Design, Fabrication, and Characterization of Muscle-Powered Biological Machines," *Nature Protocols* 12, no. 3 (2017): 519–533, https://doi.org/10.1038/nprot.2016.185.

17. Lauren Grant, Ritu Raman, Caroline Cvetkovic, Meghan C. Ferrall-Fairbanks, Gelson J. Pagan-Diaz, Pierce Hadley, Eunkyung Ko, Manu O. Platt, and Rashid Bashir, "Long-Term Cryopreservation and Revival of Tissue Engineered Skeletal Muscle," *Tissue Engineering Part A* 25, no. 13–14 (2019): 1023–1036, https://doi.org/10.1089/ten.TEA.2018.0202; Yuya Morimoto, Hiroaki Onoe, and Shoji Takeuchi, "Biohybrid Robot with Skeletal Muscle Tissue Covered with a Collagen Structure for Moving in Air," *APL Bioengineering* 4, no. 2 (2020): 026101, https://doi.org/10.1063/1.5127204.

**Chapter 7**

1. Hanna L. Tuomisto and M. Joost Teixeira De Mattos, "Environmental Impacts of Cultured Meat Production," *Environmental Science and Technology* 45, no. 14 (2011): 6117–6123, https://doi.org/10.1021/es200130u; "World Population Prospects 2019," United Nations, Population Division, 2019, https://population.un.org/wpp/Download/Standard/Population/.

2. Tom Ben-Arye and Shulamit Levenberg, "Tissue Engineering for Clean Meat Production," *Frontiers in Sustainable Food Systems* 3 (June 2019): 1–19, https://doi.org/10.3389/fsufs.2019.00046.

3. Guoqiang Zhang, Xinrui Zhao, Xueliang Li, Guocheng Du, Jingwen Zhou, and Jian Chen, "Challenges and Possibilities for Bio-Manufacturing Cultured Meat," *Trends in Food Science & Technology* 97 (2020): 443–450.

4. Liz Specht, *An Analysis of Culture Medium Costs and Production Volumes for Cultivated Meat* (Washington, DC: Good Food Institute, 2019), 1–30, https://www.gfi.org/files/sci-tech/clean-meat-production-volume-and-medium-cost.pdf.

5. Tuomisto and Joost Teixeira De Mattos, "Environmental Impacts," 6117–6123.

6. Carolyn S. Mattick, Amy E. Landis, Braden R. Allenby, and Nicholas J. Genovese, "Anticipatory Life Cycle Analysis of In Vitro Biomass Cultivation

for Cultured Meat Production in the United States," *Environmental Science and Technology* 49, no. 19 (2015): 11941–11949, https://doi.org/10.1021/acs.est .5b01614.

7. John Lynch and Raymond Pierrehumbert, "Climate Impacts of Cultured Meat and Beef Cattle," *Frontiers in Sustainable Food Systems* 3, no. 5 (February 2019): 1–11, https://doi.org/10.3389/fsufs.2019.00005.

8. Vincent Bodiou, Panagiota Moutsatsou, and Mark J. Post, "Microcarriers for Upscaling Cultured Meat Production," *Frontiers in Nutrition* 7, no. 10 (February 2020): 1–16, https://doi.org/10.3389/fnut.2020.00010.

9. Neil Stephens, Lucy Di Silvio, Illtud Dunsford, Marianne Ellis, Abigail Glencross, and Alexandra Sexton, "Bringing Cultured Meat to Market: Technical, Socio-Political, and Regulatory Challenges in Cellular Agriculture," *Trends in Food Science & Technology* 78 (June 2017): 155–166, https://doi.org/10.1016/j .tifs.2018.04.010.

**Chapter 8**

1. Peter T. Macklem and Andrew Seely, "Towards a Definition of Life," *Perspectives in Biology and Medicine* 53, no. 3 (2010): 330–340, https://doi.org/10.13 53/pbm.0.0167; Serhiy A. Tsokolov, "Why Is the Definition of Life So Elusive? Epistemological Considerations," *Astrobiology* 9, no. 4 (2009): 401–412, https://doi.org/10.1089/ast.2007.0201; Helen Greenwood Hansma, "Life = Self-Reproduction with Variations?," *Journal of Biomolecular Structure and Dynamics* 29, no. 4 (2012): 621–622, https://doi.org/10.1080/0739110120 10525007.

2. Annelien L. Bredenoord, Hans Clevers, and Juergen A. Knoblich, "Human Tissues in a Dish: The Research and Ethical Implications of Organoid Technology," *Science* 355, no. 6322 (2017): 1–7, https://doi.org/10.1126/science .aaf9414.

3. Ritu Raman and Robert Langer, "Biohybrid Design Gets Personal: New Materials for Patient-Specific Therapy," *Advanced Materials* 32, no. 13 (2019): 1–19, https://doi.org/10.1002/adma.201901969.

4. Robert D. Truog, Aaron S. Kesselheim, and Steven Joffe, "Paying Patients for Their Tissue: The Legacy of Henrietta Lacks," *Science* 336, no. 6090 (2012): 37–38, https://doi.org/10.1126/science.1216888.

5. I. A. Otto, C. C. Breugem, J. Malda, and A. L. Bredenoord, "Ethical Considerations in the Translation of Regenerative Biofabrication Technologies into Clinic and Society," *Biofabrication* 8, no. 4 (2016): 042001.

6. Judith Jarvis Thomson, "The Trolley Problem," *The Yale Law Journal* 94, no. 6 (1985): 1395–1415, https://doi.org/10.1119/1.1976413.

7.  Insoo Hyun, "Engineering Ethics and Self-Organizing Models of Human Development: Opportunities and Challenges," *Cell Stem Cell* 21, no. 6 (2017): 718–720, https://doi.org/10.1016/j.stem.2017.09.002.

8.  Jarvis Thomson, "The Trolley Problem," 1395–1415.

9.  Matthew Sample, Marion Boulicault, Caley Allen, Rashid Bashir, Insoo Hyun, Megan Levis, Caroline Lowenthal et al., "Multi-Cellular Engineered Living Systems: Building a Community around Responsible Research on Emergence," *Biofabrication* 11, no. 4 (2019): 043001, https://doi.org/10.1088/1758–5090 /ab268c.

10. "Ethics," EBICS, 2017, http://Ebics.Net/Knowledge-Transfer/Ethics.

11. Peter T. Macklem, "Emergent Phenomena and the Secrets of Life," *Journal of Applied Physiology* 104, no. 6 (2008): 1844–1846, https://doi.org/10.1152 /japplphysiol.00942.2007.VIEWPOINT.

**Chapter 9**

1.  Wendy C. Newstetter, Essy Behravesh, Nancy J. Nersessian, and Barbara B. Fasse, "Design Principles for Problem-Driven Learning Laboratories in Biomedical Engineering Education," *Annals of Biomedical Engineering* 38, no. 10 (2010): 3257–3267, https://doi.org/10.1007/s10439-010-0063-x; Jennifer R. Amos, Troy J. Vogel, and Princess Imoukhuede, "Assessing Teaming Skills and Major Identity through Collaborative Sophomore Design Projects across Disciplines," in *American Society for Engineering Education* (Washington, DC: ASEE, 2015), 2–11; Thomas R. Harris, John D. Bransford, and Sean P. Brophy, "Roles for Learning Sciences and Learning Technologies in Biomedical Engineering Education: A Review of Recent Advances," *Annual Review of Biomedical Engineering* 4, no. 1 (2002): 29–48, https://doi.org/10.1146/annurev .bioeng.4.091701.125502; Michelle C. Laplaca, Wendy C. Newstetter, and Ajit P. Yoganathan, "Problem-Based Learning in Biomedical Engineering Curricula," in *ASEE/IEEE Frontiers in Education Conference* (Washington, DC: ASEE, 2001), 16–21.

2.  David Jonassen, Johannes Strobel, and Chwee Beng Lee, "Everyday Problem Solving in Engineering: Lessons for Engineering Educators," *Journal of Engineering Education* 95, no. 2 (2006): 139–151; Clive L. Dym, Alice M. Agogino, Ozgur Eris, Daniel D. Frey, and Larry J. Leifer, "Engineering Design Thinking, Teaching, and Learning," *Journal of Engineering Education* 94, no. 1 (January 2005): 103–120; Kathryn F. Trenshaw, Jerrod A. Henderson, Marina Miletic, Edmund G. Seebauer, Ayesha S. Tillman, and Troy J. Vogel, "Integrating Team-Based Design across the Curriculum at a Large Public University," *Chemical Engineering Education* 48, no. 3 (2014): 139–148.

3. Ritu Raman, Marlon Mitchell, Pablo Perez-Pinera, Rashid Bashir, and Lizanne DeStefano, "Design and Integration of a Problem-Based Biofabrication Course into an Undergraduate Biomedical Engineering Curriculum," *Journal of Biological Engineering* 10, no. 1 (2016): 1–8, https://doi.org/10.1186/s13036-016-0032-5.

4. Jorge E. Monzon and Alvaro Monzon-Wyngaard, "Ethics and Biomedical Engineering Education: The Continual Defiance," in *Proceedings of the 31st Annual International Conference of the IEEE Engineering in Medicine and Biology Society: Engineering the Future of Biomedicine, EMBC 2009* (Piscataway, NJ: IEEE, 2009): 2011–2014, https://doi.org/10.1109/IEMBS.2009.5333435; Taylor Martin, Karen Rayne, Nate J. Kemp, and Kenneth R. Diller, "Teaching for Adaptive Expertise in Biomedical Engineering Ethics," *Science and Engineering Ethics* 11, no. 2 (2005): 257–276.

# FURTHER READING

Atala, Anthony, and James J. Yoo, eds. *Essentials of 3D Biofabrication and Translation*. Academic Press, 2015.

Baylis, Françoise. *Altered Inheritance: CRISPR and the Ethics of Human Genome Editing*. Harvard University Press, 2019.

Goldfield, Eugene C. *Bioinspired Devices*. Harvard University Press, 2018.

Hockfield, Susan. *The Age of Living Machines: How Biology Will Build the Next Technology Revolution*. WW Norton & Company, 2019.

Joachim, Mitchell, and Nina Tandon. *Super Cells: Building with Biology*. TED Books, 2014.

Kaebnick, Gregory E., and Thomas H. Murray, eds. *Synthetic Biology and Morality: Artificial Life and the Bounds of Nature*. MIT Press, 2013.

Landecker, Hannah. *Culturing Life*. Harvard University Press, 2007.

Lanza, Robert, Robert Langer, Joseph P. Vacanti, and Anthony Atala, eds. *Principles of Tissue Engineering*. Academic Press, 2020.

Lee, Suzanne, Warren Du Preez, and Nick Thornton-Jones. *Fashioning the Future: Tomorrow's Wardrobe*. Thames and Hudson, 2005.

Mukherjee, Siddhartha. *The Gene: An Intimate History*. Scribner, 2016.

Skloot, Rebecca. *The Immortal Life of Henrietta Lacks*. Broadway Paperbacks, 2017.

# INDEX

**The MIT Press Essential Knowledge Series**

RITU RAMAN, PHD, is an engineer, writer, and educator with a passion for building machines powered by biological materials that work with the human body to fight disease and damage. She received her BS *magna cum laude* from Cornell University in 2012 and her PhD from the University of Illinois at Urbana-Champaign in 2016 and trained as a postdoctoral fellow with the renowned Professor Robert Langer at the Massachusetts Institute of Technology. Ritu grew up in India, Kenya, and the United States and has championed many initiatives to increase diversity in science and use technical innovation to drive positive social change. She holds several awards and honors for both scientific innovation and outreach, including being named to the Forbes 30 Under 30 Science list and the MIT Technology Review 35 Innovators Under 35 list and receiving a L'Oréal USA Women in Science Fellowship and Ford Foundation Fellowship from the National Academies of Science, Engineering, and Medicine.